T0075251

APPLIED SYSTEM INNOVATION

PROCEEDINGS OF THE 2015 INTERNATIONAL CONFERENCE ON APPLIED SYSTEM INNOVATION (ICASI 2015), 22–27 MAY 2015, OSAKA, JAPAN

Applied System Innovation

Editors

Dr. Teen-Hang Meen
National Formosa University, Taiwan

Dr. Stephen D. Prior
The University of Southampton, UK

Dr. Artde Donald Kin-Tak Lam
Fuzhou University, China

CRC Press is an imprint of the
Taylor & Francis Group, an **informa** business
A BALKEMA BOOK

CRC Press/Balkema is an imprint of the Taylor & Francis Group, an informa business

© 2016 Taylor & Francis Group, London, UK

Typeset by V Publishing Solutions Pvt Ltd., Chennai, India
Printed and bound in Great Britain by CPI Group (UK) Ltd, Croydon, CR0 4YY

All rights reserved. No part of this publication or the information contained herein may be reproduced, stored in a retrieval system, or transmitted in any form or by any means, electronic, mechanical, by photocopying, recording or otherwise, without written prior permission from the publisher.

Although all care is taken to ensure integrity and the quality of this publication and the information herein, no responsibility is assumed by the publishers nor the author for any damage to the property or persons as a result of operation or use of this publication and/or the information contained herein.

Published by: CRC Press/Balkema
P.O. Box 11320, 2301 EH Leiden, The Netherlands
e-mail: Pub.NL@taylorandfrancis.com
www.crcpress.com – www.taylorandfrancis.com

ISBN: 978-1-138-02893-7 (Hbk + USB-card)
ISBN: 978-1-4987-8059-9 (eBook PDF)

Applied System Innovation – Meen, Prior & Lam (Eds)
© 2016 Taylor & Francis Group, London, ISBN 978-1-138-02893-7

Table of contents

Communication Science & Engineering

Computer Science & Information Technology

Management Science

Other

Applied System Innovation – Meen, Prior & Lam (Eds)
© 2016 Taylor & Francis Group, London, ISBN 978-1-138-02893-7

Preface

The 2015 International Conference on Applied System Innovation (ICASI 2015) was held in Japan May 22–27, 2015 and provided a unified communication platform for researchers in a wide area of topics. Professionals from industry, academia and government were encouraged to discourse on research and development, professional practice, business and management in the information, innovation, communication and engineering fields. This conference enables interdisciplinary collaboration between science and engineering technologists in the academic and industry fields as well as networking internationally. Attendees benefitted from various activities useful in bringing together a diverse group of engineers and technologists from across disciplines for the generation of new ideas, collaboration potential and business opportunities. The conference had received 1063 submitted papers, of which 421 papers were selected by the committees to be presented at the ICASI 2015 conference. These papers on various topics were divided into 13 Regular Sessions and 13 Invited Sessions, and were presented in several parallel sessions at the conference. The ICASI 2015 committees selected 226 best papers to publish in this book. The committees wish for this book to enable interdisciplinary collaboration between science and design technologists in the academic and industry fields as well as networking internationally.

Advanced Material Science & Engineering

Applied System Innovation – Meen, Prior & Lam (Eds)
© 2016 Taylor & Francis Group, London, ISBN 978-1-138-02893-7

Ultradrawing and ultimate tenacity properties of Ultra-High Molecular Weight Polyethylene composite fibers filled with nanosilica particles with varying specific surface areas

Jen-Taut Yeh

Hubei Collaborative Innovation Center for Advanced Organic Chemical Materials Ministry of Education Key Laboratory for the Green Preparation and Application of Functional Materials, Faculty of Materials Science and Engineering, Hubei University, Wuhan, China
Graduate School of Material Science and Engineering, National Taiwan University of Science and Technology, Taipei, Taiwan
Department of Materials Engineering, Kun Shan University, Tainan, Taiwan

Ding Yi

Hubei Collaborative Innovation Center for Advanced Organic Chemical Materials Ministry of Education Key Laboratory for the Green Preparation and Application of Functional Materials, Faculty of Materials Science and Engineering, Hubei University, Wuhan, China

Chuen-Kai Wang, Lu-Kai Huang & Chih-Chen Tsai

Graduate School of Material Science and Engineering, National Taiwan University of Science and Technology, Taipei, Taiwan

Wei-Yu Lai

Department of Materials Engineering, Kun Shan University, Tainan, Taiwan

ABSTRACT: Functionalized nanosilica ($FNSI^x_{my}$) particles were prepared by grafting maleic anhydride grafted polyethylene onto three types of nanosilica (NSI^x) particles with a quoted specific surface area of 100, 300 and 600 m^2/g, respectively. Original and/or functionalized nanosilica particles were used to investigate their influence on the ultradrawing and ultimate tensile properties of UHMWPE/nanosilica (F_{100} NSI^x_z) and UHMWPE/modified nanosilica ($F_{100}FNSI^x_{my-z}$) fibers. The achievable draw ratios (D_{ra}) of F_{100} NSI^x_z and $F_{100}FNSI^x_{my-z}$ as-prepared fibers approached a maximal value as NSI^x and/or $FNSI^x_{my}$ nanosilica contents reached an optimal value, respectively. In which, D_{ra} values of $F_{100}FNSI^{100}_{my-0.075}$, $F_{100}FNSI^{300}_{my-0.05}$ and $F_{100}FNSI^{600}_{my-0.0375}$ as-prepared fibers prepared at the optimal $FNSI^x_{my}$ contents at 0.075, 0.05 and 0.0375 phr, respectively, were significantly higher than those of the corresponding F_{100} $NSI^{100}_{0.1}$, F_{100} $NSI^{300}_{0.0625}$ and F_{100} $NSI^{600}_{0.05}$ as-prepared fibers with an optimal NSI^x content at 0.1, 0.0625 and 0.05 phr, respectively. Moreover, $F_{100}FNSI^{100}_{my-0.075}$, $F_{100}FNSI^{300}_{my-0.05}$ and $F_{100}FNSI^{600}_{my-0.0375}$ as-prepared fibers prepared at the optimal $FNSI^x_{my}$ contents exhibited the highest D_{ra} values, respectively, as their $FNSI^{100}_{my}$, $FNSI^{300}_{my}$ and $FNSI^{600}_{my}$ particles were modified using an optimal weight ratio of PE_{g-MAH} to NSI^{100}, NSI^{300} and NSI^{600} at 3, 6 and 9, respectively. The highest D_{ra} values obtained for $F_{100}FNSI^{100}_{m3-0.075}$, $F_{100}FNSI^{300}_{m6-0.05}$ and $F_{100}FNSI^{600}_{m9-0.0375}$ as-prepared fibers prepared at the optimal $FNSI^x_{my}$ contents and weight ratios of PE_{g-MAH} to NSI^x increased significantly as the specific surface areas of $FNSI^x_{my}$ particles increased. Similar to those found for their orientation factor (f_o) values, tensile strength (σ_f) and initial modulus (E) values of drawn F_{100} NSI^x_z and $F_{100}FNSI^x_{my-z}$ fiber specimens reached a maximal value as their NSI^x and/or $FNSI^x_{my}$ contents approach an optimal value, respectively. In which, the σ_f and E values of drawn $F_{100}FNSI^{100}_{my-0.075}$, $F_{100}FNSI^{300}_{my-0.05}$ and $F_{100}FNSI^{600}_{my-0.0375}$ fibers prepared at the optimal $FNSI^x_{my}$ contents, were significantly higher than those of the corresponding drawn F_{100} $NSI^{100}_{0.1}$, F_{100} $NSI^{300}_{0.0625}$ and F_{100} $NSI^{600}_{0.05}$ fibers with the same draw ratio and an optimal NSI^x content, respectively. Moreover, at a fixed draw ratio, drawn $F_{100}FNSI^{100}_{my-0.075}$, $F_{100}FNSI^{300}_{my-0.05}$ and $F_{100}FNSI^{600}_{my-0.0375}$ fibers prepared at the optimal $FNSI^x_{my}$ contents exhibited the highest σ_f and E values, as their $FNSI^{100}_{my}$, $FNSI^{300}_{my}$ and $FNSI^{600}_{my}$ particles were modified using an optimal weight ratio of PE_{g-MAH} to NSI^{100}, NSI^{300} and NSI^{600} at 3, 6 and 9, respectively. The highest σ_f and E values obtained for drawn $F_{100}FNSI^{100}_{m3-0.075}$, $F_{100}FNSI^{300}_{m6-0.05}$ and $F_{100}FNSI^{600}_{m9-0.0375}$ fibers prepared at the optimal $FNSI^x_{my}$ contents and weight ratio of PE_{g-MAH} to NSI^x increased significantly as the specific surface areas of $FNSI^x_{my}$ particles increased. In fact, the ultimate σ_f value of the best prepared $F_{100}FNSI^{600}_{m9-0.0375}$ drawn fiber prepared using one-stage drawing process at 95°C reached 7.6 GPa, which is about 1.7 and 1.5 times of those of the $F_{100}FNSI^{100}_{m3-0.075}$ and the best prepared UHMWPE/functionalized carbon nanotube drawn fiber specimens, respectively, and is about 2.3 times of that of the best prepared UHMWPE drawn fibers prepared at the same optimal UHMWPE concentration and drawing condition but without addition of any nanofiller. To understand the interesting ultradrawing and tensile properties of F_{100} NSI^x_z and $F_{100}FNSI^x_{my-z}$ fibers, thermal properties of F_{100} NSI^x_z and $F_{100}FNSI^x_{my-z}$ as-prepared fibers, fourier transform infrared, specific surface areas and transmission electron microcopic analyses of the original and functionalized nanosilica particles with varying specific surface areas were performed in this study.

Keywords: ultradrawing; nanosilica; specific surface area

Applied System Innovation – Meen, Prior & Lam (Eds)
© 2016 Taylor & Francis Group, London, ISBN 978-1-138-02893-7

Dual-mode frequency response using solidly mounted resonators

W.C. Shih, Y.C. Chen, W.T. Chang, C.Y. Wen & P.W. Ting
Department of Electrical Engineering, National Sun Yat-Sen University, Kaohsiung, Taiwan

K.S. Kao
Department of Computer and Communication, Shu-Te University, Kaohsiung, Taiwan

C.C. Cheng
Department of Electronic Engineering, De Lin Institute of Technology, Taipei, Taiwan

ABSTRACT: In order to fabricate a SMR, SiO_2/Mo are chosen as the low/high acoustic impedance materials to form the Bragg reflector, and the Aluminum nitride (AlN) is adopted as the piezoelectric layer because of its high acoustic wave velocity of 10,400 m/s. To obtain the resonant frequency of about 2.5 GHz, the specific thicknesses of Mo, SiO_2, and AlN are simulated for thin film deposition processes. In addition, to obtain optimized SMR characteristics for wireless communication applications, the deposition condition of AlN thin films is investigated. The SMR device shows a resonant frequency of 1.39 GHz (shear mode) and 2.48 GHz (longitudinal mode).

Applied System Innovation – Meen, Prior & Lam (Eds)
© 2016 Taylor & Francis Group, London, ISBN 978-1-138-02893-7

A study on the properties of ADI with two-step austenitization

Dong-Shyen Yang
Department of Mechanical Engineering, Chienkuo Technology University, Changhua County, Taiwan

Ta-Jen Peng
Digital Manufacturing Technology Department, Industrial Technology Research Institute, Hsinchu City, Taiwan

ABSTRACT: This research is aimed at exploring the influences of the two-step austenitization on the microstructures and the mechanical properties of austempered ductile irons. With the two-step austenitization, the cast irons were treated with austempering. The microstructure and the distribution of the solute component were examined with Scanning Electron Microscope (SEM) and Electron Probe Micro-Analyzer (EPMA). The results show that during the second austenitization, in the matrices of the three ductile irons, eutectoid ferrite segregated in the matrix microstructures of both FCD550 and FCD700 ausferrite phase has appeared, the tensile strength and the impact of the toughness decreased slightly, but the elongation increased significantly. A kind of fracture face mixed with characteristics of both brittle and ductile fracture faces was discovered.

Applied System Innovation – Meen, Prior & Lam (Eds)
© 2016 Taylor & Francis Group, London, ISBN 978-1-138-02893-7

Railway trial section with geocomposite positioned beneath the ballast bed for lifetime extension of track geometry

L. Horníček
Faculty of Civil Engineering, Czech Technical University in Prague, Prague, Czech Republic

M. Holý
NAUE GmbH & Co. KG, Espelkamp, Germany

P. Jasanský
The Railway Infrastructure Administration, State Organization, Prague, Czech Republic

ABSTRACT: In complex geological conditions, due to the occurrence of the so-called pumping effect and the absence of adequate sub-ballast, fine-grained soil particles are gradually pushed into the ballast on railway tracks with high traffic volumes causing the subsequent deterioration of the quality of track geometric parameters. A progressive method that prevents the ballast bed contamination with fine-grained soil from the subsoil contributing, at the same time, to increased stability of the ballast bed is the use of reinforcing or stabilization geocomposites placed under the ballast. To verify the effectiveness of this method, a test section with a geocomposite consisting of a welded biaxial geogrid and non-woven geotextile was established in the Czech Republic in 2012. The test section is subjected to long-term monitoring and compared with the adjoining railway track sections. The article summarizes the findings obtained from the establishment of the test section and its monitoring up-to-now.

Applied System Innovation – Meen, Prior & Lam (Eds)
© 2016 Taylor & Francis Group, London, *ISBN 978-1-138-02893-7*

Robust Surface-Enhanced Raman Scattering substrates with massive nanogaps derived from silver nanocubes self-assembled on a massed silver mirror via 1, 2-ethanedithiol monolayer as linkage and ultra-thin spacer

Ro-Tin Lin
Department of Materials Science and Engineering, National Cheng Kung University, Tainan, Taiwan, R.O.C.

Ten-Chin Wen, Kai-Wei Tsai & Shu-Chun Cheng
Department of Chemical Engineering, National Cheng Kung University, Tainan, Taiwan, R.O.C.

ABSTRACT: The highly sensitive Surface-Enhanced Raman Scattering (SERS) substrate, via a strong localized surface plasmon resonance from efficient nanogaps, can be derived from silver nanocubes, which is self-assembled on a silver surface with 1, 2-ethanedithiol as the linkage. The existence of nanogaps between Ag nanocubes and Ag surface was verified by X-ray spectroscopy and the frequency change of plasmon resonance was corroborated by the analysis of a UV–Vis spectrophotometer. The beautiful homogeneous distribution of nanocubes on the Ag surface can be used as an effective substrate by detecting a 10^{-9} M rhodamine 6G solution with high sensitivity (enhancement factor 2.8×10^8), high reliability (6.6% standard deviation from measurements of 20 sites), and high precision (calibration curve with 99.9% correlation coefficient).

Applied System Innovation – Meen, Prior & Lam (Eds)
© 2016 Taylor & Francis Group, London, ISBN 978-1-138-02893-7

Verification of strain gauge and geodetic measurements during long-term monitoring of Gagarin bridge in working conditions

J. Bures
Institute of Geodesy, Brno University of Technology, Brno, Czech Republic

L. Klusacek, R. Necas & J. Fixel
Institute of Concrete and Masonry Structures, Brno University of Technology, Brno, Czech Republic

ABSTRACT: The paper deals with issues of measurement used to assess behavior of structures. The integrated approach of using the strain gauge and geodetic methods enables to verify both methods and evaluate temperature dependencies of structures. Nowadays, the strain gauge and geodetic measurements are used separately. However, integration of both methods contributes to find real behavior of structures. Utilization of knowledge of measurement integration can be applied for example during long term monitoring of structures or during static load test measurements. The actual measured values of deflections and structure's deformations are mainly influenced by ambient temperature, variable temperature of the structure and by other external conditions. It is necessary to determine clear values of deflection and deformation without influence of external conditions to correctly evaluate behavior of a structure. Geodetic methods determine absolute changes related to reference system, strain gauges determine relative changes within the structure with higher accuracy. Selected interesting results of verification of strain and geodetic measurements gained from long term monitoring of bridge called Gagarin (built in 1961) in working conditions are described in this article. The thermal field of the structure was measured by thermometers and thermal cameras. The aim of the research was to integrate both methods and use them effectively to design and assess structures in building practice—mainly structures with small deformations.

Applied System Innovation – Meen, Prior & Lam (Eds)
© 2016 Taylor & Francis Group, London, ISBN 978-1-138-02893-7

Epitaxial growth of GaN on ZnO micro-rod by Plasma-Assisted Molecular Beam Epitaxy

Shou-Ting You & Ikai Lo
Department of Physics, National Sun Yat-Sen University, Kaohsiung, Taiwan, R.O.C.

Jenn-Kai Tsai
National Formosa University, Yunlin, Taiwan, R.O.C.

Cheng-Hug Shih
Department of Materials Science and Opto-Electronic Engineering, National Sun Yat-Sen University, Kaohsiung, Taiwan, R.O.C.

ABSTRACT: In this paper, we have studied the growth of GaN on ZnO micro-rod by Plasma-Assisted Molecular Beam Epitaxy (PA-MBE). The samples were characterized by Focused Ion Beam (FIB), Transmission Electronmicroscope (TEM) and polarization-dependent photoluminescence. To investigate the microstructure of GaN on the polar and non-polar surfaces of ZnO, i.e. ***c***-plane (0001) and ***M***-plane (10–10), we used FIB to prepare cross-section TEM sample for the two crystal planes. We found that the high-quality ***M***-plane GaN grown on ***M***-plane ZnO can be achieved but the quality of ***c***-plane GaN grown on ***c***-plane ZnO is not as good as the former. From the interface of GaN/ZnO of the samples, we found that the other phases including Ga_2O_3 are formed at the interface. According to the experiment results, we demonstrated that the ***M***-plane ZnO surface can be used as a substrate to grow high quality ***M***-plane GaN epi-layer.

Applied System Innovation – Meen, Prior & Lam (Eds)
© 2016 Taylor & Francis Group, London, ISBN 978-1-138-02893-7

Growth of two-dimensional ZnO nanostructures on a Si substrate and their application to field emission devices

S.J. Young, L.T. Lai & T.H. Meen
Department of Electronic Engineering, National Formosa University, Yunlin, Taiwan

Y.H. Liu
Institute of Microelectronics and Department of Electrical Engineering, National Cheng Kung University, Tainan, Taiwan

L.W. Ji
Institute of Electro-Optical and Materials Science, National Formosa University, Yunlin, Taiwan

ABSTRACT: In this study, two-dimensional ZnO nanostructures were synthesized on a Si substrate by using a simple solution-based method at room-temperature. Then, field emission devices with two-dimensional ZnO nanostructures were fabricated. The thin nanosheets that were perpendicular to the Si substrate surface were mutually interwoven into net-shaped and formed a continuous nanosheet film, with a unique surface morphology and a high surface-to-volume ratio. The diameter and length of the ZnO nanosheets were 21 nm and 1.8 μm, respectively. The turn-on electric field and enhancement factors β of the fabricated devices were 5.4 V • μm^{-1} and 1014, respectively.

Keywords: ZnO nanosheet; Field Emission

Applied System Innovation – Meen, Prior & Lam (Eds)
© 2016 Taylor & Francis Group, London, ISBN 978-1-138-02893-7

Oil and cream improvements on hair keratin

Chia-Ling Chang
Department of Chemical and Materials Engineering, National Kaohsiung University of Applied Sciences, Kaohsiung, Taiwan
Min-Hwei Junior College of Health Care Management, Tainan, Taiwan

Tsung-Han Ho
Department of Chemical and Materials Engineering, National Kaohsiung University of Applied Sciences, Kaohsiung, Taiwan

Te-Hua Fang
Department of Mechanical Engineering, National Kaohsiung University of Applied Sciences, Kaohsiung, Taiwan

ABSTRACT: In this study, we use atomic force microscopy, friction force microscopy, and Fourier-Transform Infrared Spectroscopy (FTIR) to evaluate the effectiveness of nourishing hair products. The results of our experiments on nourishing oils and hair conditioner indicate that hair milk produces an improved osmotic absorption effect. The surface roughness of the horny layer is significantly lower than when cuticle repair oil is applied, thus improving the wet ability and the softness of the hair. Furthermore, the surface topography and the interfacial structure after hair damage and improvement, observed through FTIR, exhibits changes in the wave numbers. The local mechanical properties of the hair samples are herein discussed.

Applied System Innovation – Meen, Prior & Lam (Eds)
© 2016 Taylor & Francis Group, London, ISBN 978-1-138-02893-7

Growth of titanium oxide nanotubes on Ti foil substrate by using electrochemical method

Jing Liu
School of Information Engineering, Jimei University, Xiamen, Fujian, P.R. China

Yu-Chun Chen, Yun-I Ho, Wei-Yu Lin & Cheng-Fu Yang
Department of Chemical and Materials Engineering, National University of Kaohsiung, Kaohsiung, Taiwan, R.O.C.

ABSTRACT: Vertically grown TiO_2 Nano-Tube (TNT) arrays have been studied because of their preferred reduced recombination and stronger light scattering effect in the Dye-Sensitized Solar Cell (DSSC) devices. In this study, Ti foils (99.6 wt% purity) with the dimensions $15 \times 15 \times 0.1$ mm^3 were prepared and TiO_2 nanotubes were grown on them by using electrochemical method. A naturally formed TiO_2 layer at the top of the Ti foil was removed using a 0.1% HF solution under ultrasonic waves. The Ti foils were dipped into a solution of ethylene glycol containing different concentrations of ammonium fluoride (NH_4F, 0.25 wt%, 0.5 wt%, and 0.75 wt%) and 3 vol% H_2O and then a voltage of 50 V was applied with a copper (Cu) counter electrode for 2 and 3 h. After the electrochemical process the anodized Ti foil was cleaned with acetone and alcohol. From the SEM observation, the length of TiO_2 NT arrays increased with the increase of NH_4F concentration and the length of TiO_2 NT arrays had no apparent change with the increase of anodizing time. The reason to cause those results will be discussed in the paper.

Applied System Innovation – Meen, Prior & Lam (Eds)
© 2016 Taylor & Francis Group, London, *ISBN 978-1-138-02893-7*

Effect of Li_2CO_3 addition on the microstructure and electrical properties of lead-free $(Na_{0.5}K_{0.5})NbO_3$-$Bi_{0.5}(Na_{0.90}K_{0.10})_{0.5}TiO_3$

C.H. Wang
Department of Electronic Engineering, Nan Jeon University of Technology, Tainan, Taiwan
Graduate School of Engineering Science and Technology, Nan Jeon University of Technology, Tainan, Taiwan

ABSTRACT: In this paper, $0.97(Na_{0.5}K_{0.5})NbO_3$–$0.03Bi_{0.5}(Na_{0.90}K_{0.10})_{0.5}TiO_3$ [abbreviated as 0.97NKN–0.03BNKT] with the addition of 0~0.5 wt.% Li_2CO_3 has been prepared following the conventional mixed oxide process. For 0.97NKN–0.03BNKT ceramics, the electromechanical coupling coefficients of the planar mode k_p and the thickness mode k_t reach 0.30 and 0.45, respectively. In the low Li_2CO_3 content region (≤0.1 wt.%), the decrease of dielectric loss tangent together with the enhancement of mechanical quality factor correspond well to the feature of a hard doping effect on the electrical properties. For 0.97NKN–0.03BNKT ceramics by doping 0.1 wt.% Li_2CO_3, the electromechanical coupling coefficients of the planar mode k_p and the thickness mode k_t reach 0.42 and 0.59, respectively, at the sintering of 1100°C for 3 h. The significant rise of dielectric constant and electromechanical coupling factor in the low Li_2CO_3 content region (≤ 0.1 wt.%) may be attributed to the increase of grain size and a dense microstructure.

Keywords: lead-free ceramics; piezoelectric property; dielectric property; crystal structure; sintering

Applied System Innovation – Meen, Prior & Lam (Eds)
© 2016 Taylor & Francis Group, London, ISBN 978-1-138-02893-7

A Photonic Crystal Fiber sensor based on a spherical-shape structure

W. Liu, Y. Cao & Z.G. Tong
Key Laboratory of Film Electronics and Communication Devices, Tianjin University of Technology, Tianjin, China

ABSTRACT: A Photonic Crystal Fiber (PCF) sensor based on a spherical-shape structure is proposed and fabricated. The sensor head is composed of PCF and two spherical-shape structures which are made by the single mode fiber. The Refractive Index (RI) and temperature are measured by the sensor. Experimental results show refractive index sensitivity and temperature sensitivity are 195.63 nm/RIU and 9.86 pm/°C, respectively. The sensor has advantages of compact structure and simple manufacture. It can be applied in biological and medical sensing.

Keywords: Photonic Crystal Fiber; sensor; spherical-shape structure; refractive index sensitivity; temperature sensitivity

Applied System Innovation – Meen, Prior & Lam (Eds)
© 2016 Taylor & Francis Group, London, ISBN 978-1-138-02893-7

Thermal performance evaluation of the paper honeycomb board used as insulation material

K. Kobayashi & H. Matsumoto
Department of Architecture and Civil Engineering, Toyohashi University of Technology, Aichi Prefecture, Japan

ABSTRACT: One of the most effective methods for reducing energy consumption in buildings is to install insulation material inside of the outer walls. Although general insulation material made of generally glass wool or foam polystyrene have a high thermal resistance, most of them have disadvantages such as high environmental load and health risks. This study describes a thermal performance evaluation of a paper honeycomb board used as insulation. The heat resistance of materials with different shapes, thicknesses and layers of board were investigated in the model experiment. Also, the influence of the shapes on thermal performance was presented and an optimum insulation material was proposed. To reveal thermal performance, the experiments were carried out by the Guarded Hot-Plate (GHP) method to measure the specimen's actual R value. Moreover, the values were compared with the results of a CFD simulation to establish consistency in the experiment.

Keywords: insulation material; thermal performance; GHP method; CAE analysis; honeycomb board

Communication Science & Engineering

Applied System Innovation – Meen, Prior & Lam (Eds)
© 2016 Taylor & Francis Group, London, ISBN 978-1-138-02893-7

Joint DOA and DOD estimation using real-valued implementation in bistatic MIMO radars

Ann-Chen Chang
Department of Information Technology, Ling Tung University, Taichung, Taiwan

ABSTRACT: In this paper, joint Direction of Arrival (DOA) and Direction of Departure (DOD) estimators with real-valued implementation are presented by introducing a preprocessing transformation for bistatic Multiple-Input Multiple-Output (MIMO) radars. First, a two-dimensional searching estimator by exploiting Unitary Multiple Signal Classification (UMUSIC) technique with automatic pairing is presented. Second, this paper also presents a reduced-dimension UMUSIC estimator by utilizing the characteristic of Kronecker product, which only requires a one-dimension search for DOA estimation. Following this, the angle of DOD corresponding to the estimated DOA is obtained by using a shaping DOD steering vector approach. Furthermore, the DOA and DOD pairing is given automatically.

Applied System Innovation – Meen, Prior & Lam (Eds)
© 2016 Taylor & Francis Group, London, *ISBN 978-1-138-02893-7*

A generating scheme of a side-peak-free correlation function for TMBOC(6,1,4/33) signal tracking based on cross-correlations

K. Chae, S. Woo & S. Yoon
College of Information and Communication Engineering, Sungkyunkwan University, Suwon, Korea

S. Yoo, S.Y. Kim & G.-I. Jee
Department of Electronics Engineering, Konkuk University, Seoul, Korea

H. Liu
School of Electrical Engineering and Computer Science, Oregon State University, Corvallis, OR, USA

D.-J. Yeom
Agency for Defense Development, Daejeon, Korea

ABSTRACT: Despite the benefits of Time-Multiplexed Binary Offset Carrier (TMBOC) modulation such as efficient spectrum usage and high positioning accuracy, TMBOC suffers from a problem of ambiguity in tracking caused by side-peaks on its autocorrelation. In this paper, we propose a generating scheme of a side-peak-free correlation function to resolve the ambiguity problem in the TMBOC signal tracking. Specifically, we first obtain cross-correlations based on specially designed pulses that are symmetric to each other, and then, combine them to obtain a side-peak-free correlation function. In numerical results, it is confirmed that the proposed correlation function provides a better tracking performance than the conventional schemes in terms of Tracking Error Standard Deviation (TESD).

Applied System Innovation – Meen, Prior & Lam (Eds)
© 2016 Taylor & Francis Group, London, ISBN 978-1-138-02893-7

A side-peak removal scheme for cosine-phased BOC signal tracking

K. Chae, S. Woo & S. Yoon
College of Information and Communication Engineering, Sungkyunkwan University, Suwon, Korea

S. Yoo, S.Y. Kim & G.-I. Jee
Department of Electronics Engineering, Konkuk University, Seoul, Korea

H. Liu
School of Electrical Engineering and Computer Science, Oregon State University, Corvallis, OR, USA

D.-J. Yeom
Agency for Defense Development, Daejeon, Korea

ABSTRACT: In this paper, a novel side-peak removal scheme is proposed for unambiguous cosine-phased Binary Offset Carrier (BOC) signal tracking. We first design two novel locally-generated signals, and then, we obtain cross-correlations by correlating each of the locally-generated signals and the received signal, respectively. Finally, we combine the cross-correlations to yield a correlation function with no side-peaks. The proposed scheme can remove any ambiguity in cosine-phased BOC signal tracking. In numerical results, we compare tracking performances based on Tracking Error Standard Deviation (TESD) values of the proposed and conventional correlation functions, and it is confirmed that the proposed correlation function provides a better TESD performance than the conventional correlation functions.

Applied System Innovation – Meen, Prior & Lam (Eds)
© 2016 Taylor & Francis Group, London, ISBN 978-1-138-02893-7

Enhanced utilization of resource allocation scheme for real-time traffic in LTE network

Yi-Ting Mai
Department of Sport Management, National Taiwan University of Sport, Taoyuan, Taiwan

Jeng-Yueng Chen
Department of Information Networking Technology, Hsiuping University of Science and Technology, Taichung, Taiwan

Chun-Chuan Yang, Ching-Hong Fang & Chih-Chung Hu
Department of Computer Science and Information Engineering, National Chi Nan University, Nantou, Taiwan

ABSTRACT: Wireless technology for the Fourth Generation (4G) of wireless broadband communications has been standardized in recent years in order to meet the intense demand of the public. Long Term Evolution (LTE) technology provides an easy, time-saving, and low-cost method for the deployment of 4G network infrastructure. To support multimedia services and high bandwidth data delivery, LTE MAC layer also has a QoS support with many QoS Class Indicator (QCI) levels. Based on LTE current, QCI priority, and QoS requirements in UEs, the original Max Rate or Proportionally Fair (PF) algorithm could not be able to achieve their goal due to UE's dynamic capacity with a varying Channel Quality Indicator (CQI). To provide for a better QoS service over LTE networks, this study suggests that per UE's CQI state, each Resource Block (RB) should be considered simultaneously in the LTE MAC layer resource allocation. Since the DL real capacity is dynamic due to UE's periodical CQI reporting, it is necessary to consider the CQI state in the LTE scheduling. We propose an Enhanced Utilization Resource Allocation (EURA) scheme, including two novel mechanisms that can dynamically fit UEs' CQI state. Simulation results have demonstrated that the proposed EURA scheme can save more RB capacity and improve the utilization of the RB assignment.

Applied System Innovation – Meen, Prior & Lam (Eds)
© 2016 Taylor & Francis Group, London, ISBN 978-1-138-02893-7

Congestion Control for massive Machine Type Communications in LTE Networks

Jeng-Yueng Chen
Department of Information Networking Technology, Hsiuping University of Science and Technology, Taiwan

Yi-Ting Mai
Department of Sport Management, National Taiwan University of Sport, Taiwan

ABSTRACT: With the advent of IT, more and more smart devices can detect several types of environmental conditions and send the data collected or alarm the server system automatically. Such applications, similar to wireless sensor networks are being developed for past few years. With the development of next-generation LTE wireless mobile network, SA working group also developed a related technology, namely Machine Type Communication (MTC) in 3GPP. The MTC applications developed by the 3GPP SA working group can also be operated without interaction with humans, and can transmit data directly to the servers located in the LTE core network. However, the popularity of MTC devices may extensively increase due to more and more fancy applications. This may result in LTE core network overload as this would increase the transmission demand drastically. Moreover, RACH congestion situation would become more and more serious due to limited preamble slots in the LTE system. This study attempts to use a hierarchical architecture for grouping MTC devices on the basis of their respective futures. The ACB or EAB combined with group paging mechanisms are used to reduce the probability of congestion. Furthermore, this study also develops a Device-to-Device, D2D, communication approach to further reduce the LTE system load. The performance results have shown that the proposed MTC cluster architecture can reduce system load.

Applied System Innovation – Meen, Prior & Lam (Eds)
© 2016 Taylor & Francis Group, London, ISBN 978-1-138-02893-7

An Efficient LISP Locator Evaluation scheme for traffic control in Wireless Mesh Networks

Pei-Jung Lin
Department of Computer Science and Information Engineering, Hungkuang University, Taichung, Taiwan

ABSTRACT: In this work, we design an efficient LISP Locator Evaluation (LLE) scheme for traffic control in Wireless Mesh Networks (WMNs). This LLE scheme adopts the fuzzy logic theory to generate a weight value of each LISP locator based on the bandwidth utilization, data rate and number of flows in mapping servers. A mesh router queries a LISP locator with the lowest weight value from its default mapping server. Hence, the traffic load can be distributed to different locators and then improve the network performance. In order to reduce the control overhead, the LLE scheme also uses the fuzzy logic theory to generate a cache lifetime for each LISP locator. The LLE scheme considers the traffic information update frequency and lookup latency of mapping information as input linguistic variables to generate and adjust the cache lifetime of each LISP locator. In experimental results, the proposed scheme is compared with our previous work, a k-level scheme, and the Shortest-Path (SP) scheme. The simulation results indicate that our scheme outperforms the other schemes in terms of packet delivery ratio and control overhead.

Applied System Innovation – Meen, Prior & Lam (Eds)
© 2016 Taylor & Francis Group, London, ISBN 978-1-138-02893-7

Evolution of UI's Influence

Chai Jiaxiang
Guangdong Literature and Art Vocational College, Guangzhou, Guangdong, China

Cai Jiangyu
South China Normal University, Guangzhou, Guangdong, China

Kuo-Kuang Fan
Graduate School of Design, National Yunlin University of Science and Technology, Yunlin, Taiwan

ABSTRACT: As an update version of industrial design in the era of information, the changes of UI design promote the evolution of living intelligent instrument, improve its practicability and information transfer efficiency, and turn it into an extension of the biological functions of human body. As Internet of Things (IOT), cloud computing, and other information technologies keep evolving, the core of UI design will become consolidated while the visual form will diversify, thus exerting a more evident influence on the "biological" features of living intelligent instrument. The author hopes that this study will shed some light on the future development of machine-to-machine interaction represented by intelligent appliance.

Applied System Innovation – Meen, Prior & Lam (Eds)
© 2016 Taylor & Francis Group, London, ISBN 978-1-138-02893-7

A modified grouping protocol for Wireless Sensor Networks based on SDN

Jiun-Jian Liaw, Ming-Kai Hsu, Cheng-Wei Chou & Hung-Chi Chu
Department of Information and Communication Engineering, Chaoyang University of Technology, Taichung, Taiwan, R.O.C.

ABSTRACT: Since environmental monitoring and disaster prevention can be achieved by using WSN and IoT, SDN also can be used to change the networking architecture. We can integrate WSN and SDN to make a more effective system for environmental monitoring. In WSN, developing an energy efficient method for extending the network's lifetime is an important issue. In this paper, we proposed a modified protocol to extend the lifetime of the network. The proposed method is based on the group clustering method. The results of simulation show that the proposed method gets a better performance when the BS is inside the sensing area. We suggest applying the proposed method while the BS is located inside the sensing area.

Applied System Innovation – Meen, Prior & Lam (Eds)
© 2016 Taylor & Francis Group, London, ISBN 978-1-138-02893-7

Collision-free emergency message delivery in vehicular ad hoc networks

Chia-Ho Ou, Chih-Feng Chao, Wei-Pu He & Chong-Min Gao
Department of Computer Science and Information Engineering, National Pingtung University, Taiwan

ABSTRACT: The emergency message delivery is one of essential issues in Vehicular Ad Hoc Networks (VANETs). Traditional emergency message dissemination incurs the broadcast storm problem. Several schemes have been proposed for broadcasting emergency message in VANETs. However, these schemes reduce the number of rebroadcast but still may cause message collisions. To solve the problem, this paper presents a collision-free emergency message delivery scheme for VANETs. In the proposed method, each vehicle collects beacon messages of its neighbors and creates an emergency message delivery path to transmit the emergency message to other vehicles. The proposed scheme is evaluated and compared to existing solutions by the ns-2 network simulator. Overall, the simulation results show that the proposed scheme achieves 100 percent of the message penetration rate and a lower number of emergency messages on the road.

Applied System Innovation – Meen, Prior & Lam (Eds)
© 2016 Taylor & Francis Group, London, ISBN 978-1-138-02893-7

Development of an interactive water screen using Kinect based on 3D arm movement recognition

Yu-Xiang Zhao, Min-Chueh Kan & Chin-Ju Pan
Department of Computer Science and Information Engineering, National Quemoy University, Taiwan

Chien-Hsing Chou & Hui-Ju Chen
Department of Electrical Engineering, Tamkang University, New Taipei City, Taiwan

Yi-Zeng Hsieh
Department of Management and Information Technology, Southern Taiwan University of Science and Technology, Taiwan

ABSTRACT: This paper presents an interactive water screen system using Kinect based on 3D arm movement recognition technology. The water screen was designed by combining an Arduino microcontroller board with an electronic water valve, a Kinect device was employed to capture the trajectories of user hand movement to facilitate user–device interaction. The system converts hand movement data into trajectory patterns for controlling the display on the water screen. Hand movement recognition was achieved using a moving average filter and projection filter to smooth the coordinate data in order to enhance the system's recognition efficiency and accuracy. The results indicated that the recognition efficiency was higher than 90%, both overall and for individual subjects.

Applied System Innovation – Meen, Prior & Lam (Eds)
© 2016 Taylor & Francis Group, London, ISBN 978-1-138-02893-7

Box-office statistics analysis of Zhang Yimou's films in Spanish marketing

Lu-Lu Mao
The Art Institute, Xiangtan University, Hunan, P.R. China
Faculty of Arts, The University of Leon, Castilla y León, Spain

ABSTRACT: As the second-largest film market after the USA, China's box-office sales totaled 29.6 billion RMB in 2014. However, Chinese films grossed only 1.87 billion RMB overseas. This makes the overseas box office of Chinese films an important and interesting topic. Expecting the statistic analysis of films by a certain director in certain marketing can bring some suggestion to the promotion of Chinese films abroad. This paper collected the box-office sales of all those 15 films directed by Zhang Yimou, released in Spain from 1987 to 2014. After the statistic analysis of the box office and audience numbers, this study found that: 1) different types of films have a different acceptance, but the acceptance is stable within the types; 2) the same type of film has a similar cultural discount for a Spanish audience; and 3) the Spanish audience is consistent with Chinese film critics for Chinese films. Then, through the comparison of the box office of Zhang Yimou's works in Spain and the USA, the paper explored that: 1) the box-office gap between Spanish and American marketing has been steadily narrowing in recent years; 2) different from the American audience's thinking of Chinese films as myths of martial arts and kung-fu, the Spanish audience has a more tolerant acceptance to the abundant Chinese culture, not only martial arts but also modern Chinese art and life.

Applied System Innovation – Meen, Prior & Lam (Eds)
© 2016 Taylor & Francis Group, London, ISBN 978-1-138-02893-7

Blind CFO estimation using particle swarm optimization for interleaved OFDMA Uplink

Ann-Chen Chang
Department of Information Technology, Ling Tung University, Taichung, Taiwan

ABSTRACT: This paper deals with the problem of estimating Carrier Frequency Offset (CFO) under single data block for interleaved Orthogonal Frequency Division Multiple Access (OFDMA) uplink systems. It has been shown that the searching complexity and estimating accuracy of the conventional searching-based estimators strictly depend on the number of search grids used during the search. It is time consuming and the required number of search grid is not clear to determine. As proposed in this paper, particle swarm optimization with the center-symmetric trimmed correlation matrix and orthogonal projection technique is presented for the purpose of efficient estimation. It doesn't require eigenvalue decomposition and only needs single OFDMA data block. Meanwhile, the advantage of inherent interleaved OFDMA signal structure also is exploited to conquer the problems of local optimization and the effect of ambiguous peaks for the proposed approach. Finally, several simulation results are provided for illustration and comparison.

Applied System Innovation – Meen, Prior & Lam (Eds)
© 2016 Taylor & Francis Group, London, ISBN 978-1-138-02893-7

A design of multimedia CAI material on optical fiber splicing and measurement

Y.H. Yeh, Y.E. Wu & W.C. Hsu
National Kaohsiung First University of Science and Technology, Kaohsiung, Taiwan

ABSTRACT: This paper aims to design a program of multimedia Computer Assisted Instruction (CAI) material which focuses on the optical fiber maintenance teaching. By using this program, the instructors will better their teaching performance while the learners can improve their learning effects, and it also provides the learners with great opportunities of simulation practicing on the computer before they take the skill tests. Finally, the number of learners can be increased while the training cost will be decreased.

Keywords: optical fiber equipment; interactive teaching material; CAI

Computer Science & Information Technology

Applied System Innovation – Meen, Prior & Lam (Eds)
© 2016 Taylor & Francis Group, London, ISBN 978-1-138-02893-7

Review of robust biometrics based three-factor remote user authentication scheme with key agreement

T.T. Ngo & T.Y. Choe
Department of Computer Engineering, Kumoh National Institute of Technology, Gumi-si, Republic of Korea

ABSTRACT: In 2013, Li et al. proposed a robust biometrics based three-factor remote user authentication scheme with key agreement using elliptic curve and fuzzy extractor. He claimed that while providing high-level of security and more useful functions, the scheme resolves weaknesses of An's scheme such as vulnerable to denial-of-service attack and forgery attack, and no session key agreement. However, we found that his scheme is not truly based on three factors as they claimed and cannot resist stolen smart card attack, and insider attack. Therefore, we propose a new scheme that involves adding parallel processing server smart cards as a new security factor and creating a more efficient and stronger authentication protocol based on elliptic curve cryptography and a fuzzy extractor in order to withstand various types of attacks including the above mentioned attacks, and to improve system performance.

Applied System Innovation – Meen, Prior & Lam (Eds)
© 2016 Taylor & Francis Group, London, ISBN 978-1-138-02893-7

Axiomatic approaches based on the software trustworthiness measure

H.W. Tao
State Key Laboratory of Mathematical Engineering and Advanced Computing, Zhengzhou, P.R. China

Y.X. Chen
MoE Engineering Center for Software/Hardware Co-Design Technology and Application, East China Normal University, Shanghai, P.R. China

J.M. Pang
State Key Laboratory of Mathematical Engineering and Advanced Computing, Zhengzhou, P.R. China

ABSTRACT: Due to the outstanding problems of software trustworthiness, the measurement of software trustworthiness has aroused more and more concern. However, few researches pay attention to using more rigorous approaches in the measurement of software trustworthiness and carrying out the theoretical validation of measures. Axiomatic approaches by formally defining the desirable properties of the measures for software attributes can provide precise, formal, and unambiguous terms for the quantification of software attributes. We once applied axiomatic approaches to the measured software trustworthiness, presented four properties for software trustworthiness measures and gave three measures for software trustworthiness. In this paper, we propose a new property-related user expectation to establish an improved software trustworthiness measure and validate this measure from the theory point of view by proving that it complies with the five properties presented by us. A comparative study shows that this measure is better than the popular software trustworthiness measures in terms of our properties.

Applied System Innovation – Meen, Prior & Lam (Eds)
© 2016 Taylor & Francis Group, London, ISBN 978-1-138-02893-7

3-Dimension personal identification and its applications based on Kinect

C.H. Lin, J.C. Liu & S.Y. Lin
Department of Computer Science, Tunghai University, Taichung, Taiwan

ABSTRACT: Many researchers have studied issues related to motion detection and identification. Since motion capture devices were expensive in the past, general video cameras were used to capture 2D posture information of human movements for analysis. This study uses the Kinect sensor for motion identification. Besides using 3D Kinect information for identification, finger gestures are also used to implement functions. The proposed method, with both non-contact and convenient features, consists of three stages: face recognition, skeleton recognition, and interactive command sessions. It achieves secure identity recognition and enables users to interactively execute tasks by gestures. Since 3D information of a human body cannot be easily replicated, the proposed method attains security protection and is promised for various applications in the future.

Applied System Innovation – Meen, Prior & Lam (Eds)
© 2016 Taylor & Francis Group, London, ISBN 978-1-138-02893-7

High accuracy 3-D shape measurement system based on structured light

W.T. Huang, H.L. Tseng & J.C. Dai
Department of Computer Science, Minghsin University of Science and Technology, Taiwan, R.O.C.

S.Y. Tan
Graduate Institute of Computer and Communication, National Taipei University of Technology, Taiwan, R.O.C.

C.H. Chen
Department of Management Information Systems, Central Taiwan University of Science and Technology, Taichung, Taiwan, R.O.C.

ABSTRACT: As technology matures, 3D scanning systems have been used in various fields and were asked to increase their precision. Compared with the high accuracy and high-priced laser scanning system, 3D scanning systems with the structured light algorithm is relatively simple and inexpensive. Hence, our study applies the high-resolution CCD camera to capture images, hoping to achieve the high accuracy of the laser scanner. To improve the precision, both the camera and the projector require the precise calibration. Therefore, this study is to employ the Iterative Refinement and Bundle Adjustment methods for calibrating the camera and optimizing the parameters. Moreover, we apply the local homography to calculate the pixel coordinates of the image plane. Then, we use the projector—camera system to simulate the dual stereo vision correction mode to completely correct. The reprojection error of the camera calibration, projector calibration, and stereo vision calibration is within 0.5 pixels in our design. Finally, obtained by adjusting the relative positional relationship, we use the triangulation method to calculate the positional relationship information between the object and camera. This information can later be used to reconstruct 3D graphics architecture.

Applied System Innovation – Meen, Prior & Lam (Eds)
© 2016 Taylor & Francis Group, London, ISBN 978-1-138-02893-7

An innovative RFID-embedded certificate mechanism for applying and recruiting of jobs

Yu-Yi Chen
Department of Management Information System, National Chung Hsing University, Taichung, Taiwan, R.O.C.

Jinn-Ke Jan, Yao-Jen Wang & Chuan-Chiang Huang
Department of Computer Science and Engineering, National Chung Hsing University, Taichung, Taiwan, R.O.C.

ABSTRACT: With unemployment increasing in the world, diplomas and skill certificates seem to be helpful to get a good job. Although the issuing authorities have provided many types of anti-counterfeiting technology in practice, several cases of fake certificates still occur in the world. In this paper, we develop an RFID-embedded certificate verification system with both offline and online verification mechanisms. With the help of APP, the certificate record can be automatically verified and uploaded to the employment agency's site. It is convenient for the companies and job seekers to browse the resumes with the certificate records. Furthermore, there is no privacy problem since no sensitive data is stored in the tag. The system security is more enhanced and it is reliable for the recruiting companies.

Applied System Innovation – Meen, Prior & Lam (Eds)
© 2016 Taylor & Francis Group, London, ISBN 978-1-138-02893-7

Evaluate the performance of K-means distance functions with string edit distance and the open directory project

Chun-Hsiung Tseng
Department of Information Management, Nanhua University, Chiayi, Taiwan

Yung-Hui Chen
Department of Computer Information and Network Engineering, Lunghwa University of Science and Technology, Taoyuan, Taiwan

ABSTRACT: For large scale data sources such as the Web, clustering is a practical technique to make utilizing the information simpler. Among clustering algorithms, K-means has been proven as highly efficient. However, the performance of the execution of the K-means is affected by several factors and can vary dramatically. Two important factors are the selection of initial centroids and the choice of distance functions. With regard to distance functions, despite that euclidean distance is the most frequently adopted one, the researchers still wonder whether it is possible to figure out an efficient way to evaluate their performance. First, it is difficult to define what a good clustering result is. Second, it is very challenging to determine the level of "goodness" of a clustering result. In this research, we benefit from expert intelligence by using the data recorded in the Open Directory Project, which is a human-edited Web directory. Since the directory is maintained by domain experts, the way they categorize Web resources reflect the relationships more accurately. Additionally, the researchers elaborate the edit distance algorithm to measure the differences between K-means-generated clusters and categorizes shown in Open Directory Project Web site. After experimenting process on several distance functions, at the end of this paper, the results are presented along with some discussions.

Applied System Innovation – Meen, Prior & Lam (Eds)
© 2016 Taylor & Francis Group, London, ISBN 978-1-138-02893-7

Image super-resolution using hierarchical DWT coefficient replacement

Chang-Min Chou, Chiung-Wei Huang, Wei-Lun Jian & Tsong-Pu Chien
Department of Electronic Engineering, Chien Hsin University of Science and Technology, Taoyuan, Taiwan

ABSTRACT: The objective of image up-sampling (or so called super-resolution) is to produce the correlated high-resolution image from a low-resolution image. The up-sampling techniques can be widely applied to medical treatment, astronomy, supervisory system, audio/video entertainment, and signal transmission between monitors of different resolutions. This paper proposes to produce a high-resolution image from a single low-resolution image. We developed a Discrete Wavelet Transform (DWT) based super-resolution reconstruction technique which manipulates the replacements of the wavelet coefficients to achieve this goal. After applying the discrete wavelet transform on an image, the discrete wavelet coefficients of different levels are correlated in a hierarchical nature. When the high-level wavelet coefficients adjust to enhance the image resolution, the associated low-level wavelet coefficients should be adjusted or replaced at the same time to get a better super-resolution reconstruction result. Accordingly, we propose to apply discrete wavelet transform on a low-resolution image as well as a few high-resolution sample images. Then we can match the high-level wavelet coefficients of the low-resolution images to the high-resolution sample images to search for a set of wavelet coefficients for replacement on the low-resolution image. In order to get a better super-resolution result with more detail information, once a set of suitable high-level wavelet coefficients is found for replacement, we replace the wavelet coefficients of the low-resolution image not only on the high-level coefficients, but also on the correlated low-level coefficients. We have verified our concepts by a set of experiments and got some good results. However, the experimental results are somewhat messy on the details of the output of super-resolution image. We think future designs of the matching and replacement algorithms will be helpful for achieving a better super-resolution result.

Keywords: up-sampling; super-resolution; discrete wavelet transform

Applied System Innovation – Meen, Prior & Lam (Eds)
© 2016 Taylor & Francis Group, London, ISBN 978-1-138-02893-7

A study of spherical trajectory tracking using a single camera

Hung-Chi Chu, Hao Yang & Jiun-Jian Liaw
Department of Information and Communication Engineering, Chaoyang University of Technology, Taichung, Taiwan

ABSTRACT: Due to the fast progress of image processing techniques, a variety of new applications of imaging systems have been developed such as: object tracking, height measuring, spatial measurement, and vehicle speed warning. Spherical trajectory tracking is one of the object tracking researches that can be applied to motion analysis of table tennis. The key technique of the trajectory tracking is distance (or speed) measurement. In the traditional method, it requires expensive additional equipment to complete distance measurement with high accuracy. Therefore, in this paper, we proposed an image-based spherical trajectory tracking method with low cost and high estimation accuracy. The proposed method utilized the webcam to measure the known spherical size (such as a table tennis ball). The processing steps include the background extraction, image correction, and ttrajectory tracking.

Applied System Innovation – Meen, Prior & Lam (Eds)
© 2016 Taylor & Francis Group, London, ISBN 978-1-138-02893-7

Device-free indoor localization and tracking using wireless sensor networks

Mekuanint Agegnehu Bitew, Rong-Shue Hsiao, Shinn-Jong Bair,
Hsin-Piao Lin & Ding-Bing Lin
Department of Electronic Engineering, National Taipei University of Technology, Taipei, Taiwan

ABSTRACT: Recently Device-free Localization (DfL) is receiving tremendous interests to localize a target object without attacfhing any device to it. Most DfL systems are based on the fact that Received Signal Strength Indicator (RSSI) is affected by the presence and movement of people in the monitored environment. But the RSSI information is not always reliable, especially in indoor environments, as a slight environmental changes can cause a significant fluctuation of RSSI. In addition, the accuracy of RSSI-based DfL systems decrease as the distance between the transmitter and receiver increases. To address these issues, we proposed a two phase hybrid localization scheme which uses Radio Frequency (RF) and infrared signals. In the first phase, region-based passive radio map of the monitored area is created. In the second phase, the region of the target person is identified by Pyroelectric Infrared (PIR) sensors and k-Nearest Neighbor algorithm is used to calibration points which are in the identified region. Experimental results showed that the 1.3 meters accuracy of the proposed method achieves 97% while PIR and RF schemes achieve 56% and 17%, respectively.

Applied System Innovation – Meen, Prior & Lam (Eds)
© 2016 Taylor & Francis Group, London, ISBN 978-1-138-02893-7

High-resolution digital camera for light field detection of low power LED with secondary optical lens

H.L. Tseng & W.T. Huang
Department of Computer Science, Minghsin University of Science and Technology, Taiwan, R.O.C.

S.Y. Tan & W.L. Chang
Graduate Institute of Computer and Communication, National Taipei University of Technology, Taiwan, R.O.C.

ABSTRACT: LED-encapsulated secondary lens increases optical field uniformity at 24% and increases brightness at 23.29%. However, the encapsulating secondary lens constantly results in nonuniform optical fields. This problem can be identified by examining the alignment of the secondary lens. Costly equipments, such as laser collimator, spectrometer, and integrating sphere, can be used to determine whether an LED is a defective product or not. However, they are too expensive. To solve this problem, we design a device, GEN2, to measure optical field uniformity by adopting a 12-bit color depth CCD sensor (to capture images) and FPGA platform (to process images). A 10-M pixel CMOS digital camera, Panasonic DMC-G3, and a 10-M pixel CCD digital camera, Sony α 350, are used in our experiment as our control group. Images are captured in RAW format and converted to grayscale before testing. On SNR test, GEN2 outperforms the other two cameras on block 22 through block 24 of Munsell colorchecker 24, which indicates that GEN2 produces less noise. DMC-G3 and α 350 have better resolving power because they are high-level digital cameras (therefore far more expensive than GEN2). SFR measurement shows that GEN2's performance is well enough. We also build a linear stage which is used to simulate deviation of secondary lens. That deviation results in a nonuniform optical field which is needed in our experiment. We experiment on GEN2, DMC-G3, and α 350 at deviations of 0 mm, 0.5 mm, 1 mm, and 1.5 mm. Output data are gathered and plotted as charts. The results show that GEN2 has a better discrimination ratio than DMC-G3 and α 350 due to GEN2 has a 12-bit color depth sensor while DMC-G3 and α 350 have merely 8-bit color depth sensors.

Applied System Innovation – Meen, Prior & Lam (Eds)
© 2016 Taylor & Francis Group, London, ISBN 978-1-138-02893-7

Enhanced document clustering using Wikipedia-based document representation

Ki-Joo Hong, Ga-Hui Lee & Han-Joon Kim
School of Electrical and Computer Engineering, University of Seoul, Seoul, Korea

ABSTRACT: Most traditional clustering methods are based on the Vector Space Model (VSM) using 'Bag of Words' (BOW) representation. However, the BOW representation which only accounts for term frequency is quite limited because it ignores semantic relations among indexed terms. To resolve this problem, this paper proposes a new method of constructing the matrices of document representation by utilizing the Wikipedia encyclopedia, with not depending on traditional VSM, to significantly enhance the quality of document clustering. Through extensive experiments with popular 20 Newsgroup dataset, we show that our proposed method notably improves clustering performance compared with the traditional VSM-based clustering method.

Applied System Innovation – Meen, Prior & Lam (Eds)
© 2016 Taylor & Francis Group, London, ISBN 978-1-138-02893-7

Building concept graphs using Wikipedia

Ga-Hui Lee, Ki-Joo Hong & Han-Joon Kim
School of Electrical and Computer Engineering, University of Seoul, Seoul, Korea

ABSTRACT: This paper proposes a novel way of automatically building a concept graph containing 'ISA' and 'ASSO' relationships by probabilistically analyzing connected hyperlinks within the Wikipedia articles. The concept graph can be built up by connecting four types of concept pairs: *relational concept pairs, infobox concept pairs, category concept pairs* and *synopsis anchor concept pairs.* The 'ISA' relationship of concept pairs can be determined by computing the subsumption probabilities between incoming links of upper concepts and outgoing links of lower concepts, which is internally represented as a partial ordering matrix. If the difference of subsumption probabilities for two concepts is small, then such a concept pair allows to define the 'ASSO' relationship. Our prototype system can produce a highly reasonable concept graph that contains not only noun-level concepts but also proper noun-level concepts form the Wikipedia articles. We confirm that the concept graph can be used as a knowledge base for developing various types of text applications.

Applied System Innovation – Meen, Prior & Lam (Eds)
© 2016 Taylor & Francis Group, London, ISBN 978-1-138-02893-7

A hybrid cloud model for a multicriteria group decision-making process

Ting-Cheng Chang & Hui Wang
Department of Computer Science Engineering, Ningde Normal University, Ningde, Fujian, China

ABSTRACT: The cloud model can qualitatively describe the concepts of randomness and fuzziness; it can resolve uncertainty in the conversion between the quality and quantity and gives a good solution to the problem of natural language in a multicriterion decision-making group. This paper refers to the expert system of the decision tree directed at issues such as cost evaluation of the multicriteria group, the decision directed for uncertain languages, defining weights, evaluation of subjectivity in expert evaluation, and proposing a decision-making method by combining the decision tree model and the cloud model. The procedures of this method are as follows: first, a large number of samples are selected for training to master the rules of the expert evaluation, followed by entering the candidate information into the appropriate integrated evaluation cloud. Then the floating cloud method is employed to gather the preferences of each of the attributes. Finally, the case set order is obtained by scoring each integrated evaluation cloud. The example analysis showed that the decision-making method displays simplicity, effectiveness, and feasibility.

Keywords: decision tree; multicriteria decision-making; integrated cloud; uncertainty

Applied System Innovation – Meen, Prior & Lam (Eds)
© 2016 Taylor & Francis Group, London, *ISBN 978-1-138-02893-7*

Routing-aware power saving for IoT networks

Yang-Hsin Fan
Department of Computer Science and Information Engineering, National Taitung University, Taitung, Taiwan

ABSTRACT: In order to form the Internet of Things (IoT), more and more devices are implemented with the ability for surfing the internet. Those devices include PCs, smartphones, wearables, and so on. Recently, gathering these devices it resulted in co-operative smart homes, smart offices, and smart cities that are continually being developed. However, it is important to do power saving for the huge smart devices that work throughout the year. This study proposes a routing-aware power saving for IoT networks. The routing-aware rule depends on the order of devices in IoT networks. Experimental results achieve the power saving by determining the appropriate sensor system benchmarks.

Applied System Innovation – Meen, Prior & Lam (Eds)
© 2016 Taylor & Francis Group, London, ISBN 978-1-138-02893-7

Detection of multiple moving targets on the ground based on aerial imaging

Chao-Ho Chen, Tsong-Yi Chen & Zai-Ci Jiao
Department of Electronic Engineering, National Kaohsiung University of Applied Sciences, Kaohsiung, Taiwan, R.O.C.

ABSTRACT: In this research, we developed a detection technique for multiple moving targets (e.g., people and/or vehicles) on the ground from an aerial video. The proposed method is mainly composed of three parts. (1) Classification of the feature points: we first need to search the feature points at each frame using a motion vector, and each feature point is classified into a moving-object area or the background by the Delaunay triangulation technique. (2) Generation of moving-object mask: the motion vectors of the feature points in the background are employed for deriving a conversion matrix by an affine transformation and such a conversion matrix is used to obtain the background deformation that is used to form the moving-object's contour for generating a moving-object mask. (3) Labeling and tracking of the moving-object: from the above masks of the moving-objects, each moving-object is labeled and tracked. The early experimental results show that the proposed technique can effectively detect and track the moving-objects on the ground in real-time.

Applied System Innovation – Meen, Prior & Lam (Eds)
© 2016 Taylor & Francis Group, London, *ISBN 978-1-138-02893-7*

Going deep: Improving Music Emotion Recognition with layers of Support Vector Machines

Yu-Jen Hsu & Chia-Ping Chen
Department of Computer Science and Engineering, National Sun Yat-sen University, Kaohsiung, Taiwan

ABSTRACT: We propose an automatic recognition architecture which we call Deep Support Vector Machines (DSVM). Essentially, the signal processing for pattern recognition to extract salient features from the data is facilitated by the machine learning instrument of support vector machines. In a DSVM, layers of SVMs are organized in a self-repeating manner. Furthermore, the output values of an SVM at a given layer are combined with the input acoustic features to form the input vector of the SVM at the next layer. The self-repeating organization continues for several layers until the performance converges. For evaluation, we prepare a data set of 525 music clips, which are 10 seconds long and labeled with emotion categories for music emotion recognition tasks. The experimental results clearly show that the recognition accuracy improves as the number of layers in DSVM increases.

Keywords: deep learning; music emotion recognition; support vector machine

Applied System Innovation – Meen, Prior & Lam (Eds)
© 2016 Taylor & Francis Group, London, ISBN 978-1-138-02893-7

Modified Census Transform using Haar wavelet transform

Shih-Cian Huang, Jiun-Jian Liaw & Hung-Chi Chu
Department of Information and Communication Engineering, Chaoyang University of Technology, Taichung, Taiwan

ABSTRACT: In this paper, we proposed a Haar wavelet method to improve Census Transform. We applied the traditional Census Transform with the low-band Haar wavelet. The high-band Haar wavelet was used to modify the disparity, and the low-value (low-frequency) pixel got to be exerted to modify the disparity in the high-band information. It was started from the position of the low-value pixel, and then move up, down, left, and right. Each path was stopped when the pixel value was high (high frequency). Disparities of pixels in all paths were collected, so we could obtain the histogram of the disparities, in which the maximum number was set as the new disparity of the pixel. In the experiment, we have proved that our method is helpful for stereo correspondence problems, and the results have confirmed that our method can reduce the transform point and improve the accuracy in a small transform window, and improves Stereo Vision System to work faster and with better accuracy to find the correct disparity.

Applied System Innovation – Meen, Prior & Lam (Eds)
© 2016 Taylor & Francis Group, London, ISBN 978-1-138-02893-7

Digital watermarking by using orthogonal line screens

Yu-Lan Chiu, Hsi-Chun Wang & Yi-Ting Tsai
Department of Graphic Arts and Communications, National Taiwan Normal University, Taipei, Taiwan

ABSTRACT: With the rapid growth of information technology, more and more sophisticated machines for image duplication have been created to enable image reproduction in an efficient and convenient way. However, it raises a serious threat on intellectual property rights. The purpose of this study is to compose watermarks with orthogonal lines in an effort to protect copyright and to ensure anti-counterfeiting protection. To design the orthogonal line screen watermarks, vertical and horizontal threshold matrix are processed by digital halftoning and the effect of encryption and decryption on the printout of the anti-counterfeiting watermarks was evaluated. The results showed that conducting orthogonal lines to encrypt messages in documents can be successfully carried out. This technique is also applied to encrypt answers of the problem sets in text books.

Applied System Innovation – Meen, Prior & Lam (Eds)
© 2016 Taylor & Francis Group, London, *ISBN 978-1-138-02893-7*

Improvement of post-process system using vector quantization on harmonic center frequency to regenerate vowel spectrum for speech enhancement

Ching-Ta Lu
Department of Information Communication, Asia University, Taichung City, Taiwan, R.O.C.

Kun-Fu Tseng & Chih-Tsung Chen
Department of Multimedia and Game Science, Asia-Pacific Institute of Creativity, Miaoli County, Taiwan, R.O.C.

ABSTRACT: Most speech enhancement systems suffer from musical residual noise and speech distortion. This study attempts to improve the performance of a speech enhancement system by more reduction on musical residual noise; while speech quality can be also improved. Initially, background noise is efficiently removed by a speech enhancement system. A codebook for harmonic center frequency is trained by using clean speech. In the reconstruction phase, the center frequencies of robust harmonics is estimated and employed to select the best codeword from the trained harmonic codebook. An adaptive comb filter is employed to regenerate harmonic spectra, enabling the harmonic spectra of pre-processed speech to be boosted to improve speech quality for a vowel. Moreover, the harmonic compensated speech is processed by a two-dimension spectrogram filter to reduce the quantity of musical residual noise, enabling the post-processed speech to sound less annoying. Consequently, the quality of enhanced speech can be efficiently improved.

Applied System Innovation – Meen, Prior & Lam (Eds)
© 2016 Taylor & Francis Group, London, *ISBN 978-1-138-02893-7*

Analysis of security framework and protocol supporting for Internet of Things

Jung Tae Kim T
Mokwon University, Korea

ABSTRACT: The Internet of Things (IoT) refers to the uniquely identifiable objects that can interact with other objects through the medium of wireless and wire communication. Connecting the small devices to the terminal of Internet, the devices such as RFID, NFC, and sensor nodes can be identified and tracked, and can monitor the objects attached with tags in real time. This is so-called Internet of Things (IoT). A sensor is often regarded as a prerequisite device for the IoT. In this paper we introduced and surveyed the critical issues for technologies and securities of the IoT and discussed the applications and challenges of a home network related to the IoT systems.

Applied System Innovation – Meen, Prior & Lam (Eds)
© 2016 Taylor & Francis Group, London, ISBN 978-1-138-02893-7

Applying multivariate analysis of variance and Kansei engineering theory to the website page design method

Chen-Yuan Liu & Pei-Yu Liao
Department of Visual Communication Design, National Yunlin University of Science and Technology, Douliu, Taiwan

ABSTRACT: In recent years, electronic commerce is considered as the interface for presenting information to customers, which is mostly focused on human perception in terms of users' comprehension and mental demands. Thus, the visual effect of website page layout is especially important in electronic commerce interface design and assumes a marketing strategy to meet customer needs. With the prosperous growth of electronic commerce, the cascading style sheets framework supplies designers with the structure of website and modeling template, website towards cross-platform and can be used extensively in other web devices. This paper aims to derive website page styles from design factors of essential web framework and accord with the demands of the market, and pursue to provide a model for web page image design that meets customer needs. First, the semantic differential technique is used to investigate the influence of visual elements on website banner and content framework and then Kansei engineering analysis of visual components properties of cognitive is presented. Finally, multivariate analysis of variance method is used to select an optimal design strategy. The research presents a visual interface design approach for web developers to reduce cost, and proposes an optimal decision to deal with a website image design. The results of website page design model are applicable to other forms of cross-platform interface design that contributes to the electronic commerce website design.

Keywords: web banner, web page image design, Kansei engineering method, cascading style sheets, electronic commerce, multivariate analysis of variance (MANOVA)

Applied System Innovation – Meen, Prior & Lam (Eds)
© 2016 Taylor & Francis Group, London, ISBN 978-1-138-02893-7

An HSAIL ISA conformed GPU platform

Heng-Yi Chen, Chung-Ho Chen, Yun-Chi Huang, & Kuan-Chieh Hsu
Department of Electrical Engineering and Institute of Computer and Communication Engineering, National Cheng Kung University, Tainan, Taiwan

Chen-Chieh Wang
Industrial Technology Research Institute, Hsinchu, Taiwan

ABSTRACT: This paper presents a GPU platform based on the HSA (Heterogeneous System Architecture). The platform contains the OpenCL programming interface, a cycle-level HSAIL GPU simulator, and a finalizer that is able to translate the HSAIL code to our custom binary ISA. The system platform is able to run OpenCL applications, and the results are verified with the real GPU hardware. To improve the scheduling efficiency, we explore warp scheduling policies on our platform including Fair Round-Robin, Loose Round- Robin, Two-Level, and Greedy-Then-Oldest. We found that these algorithms are not optimal in using the load/store units. We propose a Memory Access First (MAF) mechanism which can be adopted to the Two- Level and Greedy-Then-Oldest scheduling policies to reduce the idle time of the load/store units. With the proposed mechanism, the warp scheduler can remove 23 to 34 percent of load/store unit idle time and improve the GPU performance by about 16%.

Applied System Innovation – Meen, Prior & Lam (Eds)
© 2016 Taylor & Francis Group, London, ISBN 978-1-138-02893-7

FPGA implemented architecture for spike sorting based on the Generalized Hebbian Algorithm

Chien-Min Ou
Department of Electronic Engineering, Chien Hsin University of Science and Technology, Taiwan

Jih-Fu Tu
Department of Electronic Engineering, St. John's University, Taiwan

Hao Chen
Department of Computer Science and Information Engineering, National Taiwan Normal University, Taiwan

ABSTRACT: Efficient VLSI architecture for fast spike sorting that can perform feature extraction based on the Generalized Hebbian Algorithm (GHA) is presented. The GHA allows efficient computation of principal components for subsequent clustering and classification operations. The GHA hardware implementation features high throughput and classification success rates. The proposed architecture is implemented by a field programmable gate array, which is embedded in a system-on-programmable-chip platform for performance measurement. Experimental results indicate that the proposed architecture is an efficient spike sorting design with high-speed computation for spike trains corrupted by significant noise.

Applied System Innovation – Meen, Prior & Lam (Eds)
© 2016 Taylor & Francis Group, London, ISBN 978-1-138-02893-7

A study of interactive navigation in artifact collection agencies

Ding-Yu Liu
Department of Applied Geoinformatics, Chia Nan University of Pharmacy and Science, Tainan City, Taiwan

Li-Zone Chang
Department of Tourism and Management, Chia Nan University of Pharmacy and Science, Tainan City, Taiwan

Kuei-Shu Hsu, Ling-Feng Lai & Chi-Wun Yang
Department of Applied Geoinformatics, Chia Nan University of Pharmacy and Science, Tainan City, Taiwan

ABSTRACT: With the advances in information technology, artifact collection agencies have also begun to use information technology, wireless mobile devices, and digital content to build an interactive navigation environment. In the past, the museum navigation methods were mostly lacking a functional interaction between users. However, in recent years, with the use of smart phones and the popularity of mobile devices, the exhibition of artifact collection mechanisms gradually established a new way of interaction patterns, even joined by virtual reality elements.

Therefore, this study focuses on artifact collection agencies' navigation forms, using the Technology Acceptance Model and questionnaires, to understand how the artifact collection agencies provide interactive navigation, learn visitors' acceptance, and make relevant recommendations. Finally, this study further proposes that in the future interactive navigation of design and a bulk of the recommendations should be provided.

Keywords: interactive navigation; virtual reality

Applied System Innovation – Meen, Prior & Lam (Eds)
© 2016 Taylor & Francis Group, London, ISBN 978-1-138-02893-7

An integration of smartphone APP with computer-aided design for manufacturing customized insole

T. Kaewwichit, H.H. Leng, J.H. Cai & C.C. Chang
Graduate Institute of Mechatronic System Engineering, National University of Tainan, Tainan, Taiwan

ABSTRACT: A foot insole plays an important factor in our foot health and walking balance. It should be carefully designed to fit the foot arch properly. It is used primarily to provide stable alignment and improvement on the body posture. Unfortunately, most of the insole manufacturers produce on a mass production basis. This often results in a poor fit because a single size does not fit all. Producing a quality insole requires clinical expertise and a proper measurement system, only available in a foot care center or in laboratories. As smartphones and computer tablets become an integral part of our life, a software Application (APP) was developed to calculate the Arch Index (AI) and to evaluate the plantar pressure distribution. The software integrates withimage-based plantar pressure measurement with a computer-aided design to customize an insole for a better fit for each individual. The advantage of the system is the flexibility of using a plantar image from several kinds of image-based plantar pressure measurement devices to calculate the AI and plantar pressure distribution. This convenient feature allows a user to analyze and obtain a tighter fit for his/her insole. The customized insole is then designed based on the AI and plantar pressure distribution targeting to change the foot structure and improve the walking balance through the usage of a 3D printer.

Keywords: customized insole; plantar pressure; arch index; smartphone APP

Computational Science & Engineering

Applied System Innovation – Meen, Prior & Lam (Eds)
© 2016 Taylor & Francis Group, London, ISBN 978-1-138-02893-7

An implicit Lie-group iterative scheme for solving the sinh-Gordon equation

C.-W. Chang
Cloud Computing and System Integration Division, National Center for High-Performance Computing, Taichung, Taiwan

ABSTRACT: In this article, the sinh-Gordon equations are coped with by contemplating the semi-discretization method and then, the resulting ordinary differential equations at the discretized spaces are numerically integrated towards the time direction by utilizing the proposed scheme to find the unknown wave quantity. When the numerical experiment is examined, we reveal that the present implicit Lie-group iterative scheme is applicable to the sinh-Gordon equation and convergent fast at each time step; moreover, we can attain the accurate, stable and effective results.

Applied System Innovation – Meen, Prior & Lam (Eds)
© 2016 Taylor & Francis Group, London, ISBN 978-1-138-02893-7

A novel vertical handover scheme for LTE-A mobile relay systems

Jeng-Yueng Chen
Department of Information Networking Technology, Hsiuping University of Science and Technology, Taichung, Taiwan

Yi-Ting Mai
Department of Sport Management, National Taiwan University of Sport, Taoyuan, Taiwan

Fongray Frank Young
Department of Communications Engineering, Feng Chia University, Taichung, Taiwan

ABSTRACT: LTE is one of 3GPP standards developed as a fourth-generation mobile communications standard. In order to solve signal strength problems when the mobile users as UEs are located on the edge of a LTE eNB signal converge area. A new entity called Relay Node (RN) was introduced in LTE-Advanced by 3GPP. Compared to eNB device, the RN has low-power consuming, low-cost features, and can be used to enlarge the eNB radio signal coverage. This manuscript focuses on RN operating in High Speed Rail (HSR) to serve high-speed rail passengers. Since the RN equipped in HSR carriage becomes moveable, we called this type of moveable RN as Mobile RN (MRN). When UEs get on the rail carriage, UEs should automatically re-setup their connections to HSR's MRN. However, the period of conventional handover procedure might not enough for leaving UEs, since there are only 2–3 minutes for a rail arrival and departure at HSR station. Also, a large number of handover procedures would be triggered by many leaving UEs at HSR station due to sudden disconnections after the rail departure. It would further instantly increase the loads of handover signal process on LTE core network. This manuscript proposes a vertical handover scheme based on prediction of the UE's movement on a HSR train. Furthermore, the proposed scheme with two mechanisms can prepare vertical handover in a short period and reduce huge signals overhead. The simulation results have demonstrated that the proposed scheme with two mechanisms has better performance than the conventional handover scheme. The proposed scheme also can smooth the handover overhead and reduce system load.

Applied System Innovation – Meen, Prior & Lam (Eds)
© 2016 Taylor & Francis Group, London, *ISBN 978-1-138-02893-7*

Design and implementation of a speech controlled omnidirectional mobile robot using a DTW-based recognition algorithm

Pi-Yun Chen, Neng-Sheng Pai, Guan-Yu Chen & Hua-Jui Kuang
Department of Electrical Engineering, National Chin-Yi University of Technology, Taichung, Taiwan

ABSTRACT: This paper presents a voice controlled omnidirectional mobile robot using a Dynamic Time Warping (DTW)-based Mandarin speech recognition algorithm. Acoustic models are built with two sets of characteristic parameters, i.e. Linear Prediction Cepstral Coefficients (LPCC) and delta cepstral coefficients, derived from Linear Predictive Coefficients (LPC) present in a pre-processed speech signal. DTW-based speech recognition is then realized by means of a comparison(s) between the built acoustic models and an input speech signal. Successful Operations of the presented voice controlled mobile robot are demonstrated at the end of this work.

Applied System Innovation – Meen, Prior & Lam (Eds)
© 2016 Taylor & Francis Group, London, ISBN 978-1-138-02893-7

Simulation of cutting process by a coated cutting tool

Chun-Mu Wu & Chen Tsai Huang
Department of Mechanical and Automation Engineering, Kao Yuan University, Kaohsiung City, Taiwan, R.O.C.

ABSTRACT: The coated cutting tool is one of the tools that is widely applied in the modern machining industry. A cutting tool coating can stabilize the chemical properties of cutting tool materials, reduce the friction coefficient in machining, and effectively reduce the heat transferred into the cutting tool. It is important and practical to research the coated cutting tool, cutting heat transfer, and cutting temperature. In the metal cutting process, the cutting temperature and tool wear caused by the cutting force are the main indicators reflecting the cut. In this paper, Deform-3D software is used to simulate the cutting process by coated cutting tool TICN and non-coated cutting tool to a C-Cr-Mo alloy (SCM440) workpiece. The cutting temperature of the tool, and workpiece, as well as the distribution situation of cutting force, and tool nose are analyzed. The results show that the coated cutting tool can effectively reduce the tool cutting temperature and improve the tool service life.

Keywords: coated cutting tool; cutting temperature; cutting force; Deform-3D

Applied System Innovation – Meen, Prior & Lam (Eds)
© 2016 Taylor & Francis Group, London, ISBN 978-1-138-02893-7

Simulation analysis of parameters of self-drilling screws

Chun-Mu Wu & Chen-Yu Lai
Department of Mechanical and Automation Engineering, Kao Yuan University, Kaohsiung, Taiwan, R.O.C.

ABSTRACT: Currently, self-drilling screw designs in the industry still rely on practical experience. However, using practical experience for product development is ineffective in both time and cost. Therefore, to achieve an optimal work efficiency, this study used a self-drilling screw with a diameter of 5 mm and a length of 28 mm to drill a steel plate for parametric simulation analysis. The finite element method of the DEFORM-3D software was used to simulate the rotation speed and feed rate parameters in the drilling process. A self-drill screw made of AISI 316H and a steel plate made of AISI 1010 were used. The self-drilling screw and steel plate were compared in terms of temperature field, stress field, and torque. The self-drilling screw's edge wearing difference was analyzed. The results showed that, in the drilling and cutting process of the self-drilling screw, different main shaft rotating speeds have insignificant influence on the equivalent stress applied to the steel plate. A high feed rate increases the processing temperature, and a high rotating speed leads to a high wearing rate.

Keywords: DEFORM-3D; stress field; temperature field; tool wearing

Applied System Innovation – Meen, Prior & Lam (Eds)
© 2016 Taylor & Francis Group, London, ISBN 978-1-138-02893-7

Numerical simulation on Friction Stir Welding of aluminum alloy and oxygen-free copper

Chun-Mu Wu & Chang-Rong Huang
Department of Mechanical and Automation Engineering, Kao Yuan University, Kaohsiung City, Taiwan, R.O.C.

ABSTRACT: Friction Stir Welding (FSW), as a new trend of solid-state welding, is one of the common welding approaches in mechanical structure welding. In particular, it is an important topic to weld high-strength weld faces on different alloys. In this paper, oxygen-free copper is subject to FSW with aluminum alloys 2024 and 6061, and the FSW formation process is simulated with software DEFORM-3D, in order to analyze the temperature field, effective strain, and flow direction of different materials in the FSW process. According to the simulation result, aluminum alloys 2024 and 6061 can be used to finish the FSW of oxygen-free copper. In addition, the author also analyzes the differences of two different metals regarding FSW optimization factor and welding strength.

Keywords: FSW; temperature field; force field; DEFORM-3D; aluminum alloy; oxygen-free copper

Applied System Innovation – Meen, Prior & Lam (Eds)
© 2016 Taylor & Francis Group, London, ISBN 978-1-138-02893-7

Quick implementation of FPGA chip control for a self-propelled vehicle searching light on the Simulink platform

C.S. Shieh
Department of Electrical Engineering, Far-East University, Hsin-Shih Tainan, Taiwan, R.O.C.

ABSTRACT: The Field Programmable Gate Array (FPGA) logic circuit provides a large number of logic gates to implement the complex circuitry on a single chip for engineers to design the desired functions. It can be used in the intelligent control design of the robot for characteristics such high capacity, low power consumption, high security, and repeatable burn; however, it is generally designed in the VHDL/Verilog language to complete the design. When the programs are more complex, its corresponding debug and simulation become more difficult in the VHDL/Verilog language platform. This paper uses the MATLAB/Simulink graphical language, and a top-down design process, not only for quick computer simulation and verification, but also for the FPGA chip to be realized immediately. The design flow of a light-sensitive navigational vehicle is proposed in the experiment to verify the feasibility of the proposed design flow.

Keywords: FPGA; MATLAB/Simulink; VHDL/Verilog language

Electrical & Electronic Engineering

Applied System Innovation – Meen, Prior & Lam (Eds)
© 2016 Taylor & Francis Group, London, ISBN 978-1-138-02893-7

Stochastic harmonic distortion and optimal filter design of MRT systems using Immune Algorithm

Hui-Jen Chuang, Wen-Yuan Tsai & Shun-Li Su
Department of Electrical Engineering, Kao Yuan University, Kaohsiung, Taiwan

Chao-Shun Chen
Department of Electrical Engineering, I-Shou University, Kaohsiung, Taiwan

Cheng-Ting Hsu
Department of Electrical Engineering, Southern Taiwan University of Science and Technology, Tainan, Taiwan

Chia-Hung Lin
Department of Electrical Engineering, National Kaohsiung University of Applied Sciences, Kaohsiung, Taiwan

ABSTRACT: The optimal design of passive filters is solved by minimising the equivalent cost of total harmonic distortion and investment cost of passive filters for mass rapid transit power systems with Immune Algorithm (IA). The objective function and constraints are expressed as antigens, and all feasible solutions are expressed as antibodies in the IA simulation process. The diversity of antibodies is then enhanced by considering the proximity of antigens so that the global optimisation during the solution process can be obtained. It is found that harmonic distortion, which results from the stochastic harmonic load flow analysis, can be mitigated effectively by the optimal planning of passive filters using the proposed immune algorithm.

Applied System Innovation – Meen, Prior & Lam (Eds)
*© 2016 Taylor & Francis Group, London,**ISBN 978-1-138-02893-7*

Fuzzy-Neural-Network Inherited Total Sliding-Mode Control for robot manipulator

Rong-Jong Wai
Department of Electronic and Computer Engineering, National Taiwan University of Science and Technology, Taipei, Taiwan

Hao-Yu Ting & Chen-Hsien Yu
Department of Electrical Engineering, Yuan Ze University, Chung Li, Taiwan

ABSTRACT: This study presents the design and analysis of an intelligent control system that inherits the robust properties of Total Sliding-Mode Control (TSMC) for an n-link robot manipulator including actuator dynamics in order to achieve a high-precision position tracking with a firm robustness. First, the coupled higher-order dynamic model of an n-link robot manipulator is introduced briefly. Then, a TSMC scheme is developed for the joint position tracking of the robot manipulator. Moreover, a Fuzzy-Neural-Network Inherited TSMC (FNNITSMC) scheme is proposed to relax the requirement of detailed system information and deal with chattering control efforts in the TSMC system. In the FNNITSMC strategy, the FNN framework is designed to mimic the TSMC law, and adaptive tuning algorithms for network parameters are derived in the sense of projection algorithm and Lyapunov stability theorem to ensure the network convergence as well as stable control performance. Experimental results of a two-link robot manipulator actuated by DC servo motors are provided to justify the claims of the proposed FNNITSMC system, and the superiority of the proposed FNNITSMC scheme is also evaluated by quantitative comparison with previous intelligent control schemes.

Applied System Innovation – Meen, Prior & Lam (Eds)
© 2016 Taylor & Francis Group, London, ISBN 978-1-138-02893-7

An inverter-based wide-band Low-Noise Amplifier with inductor peaking technique

San-Fu Wang, Chi-Kuei Chung & Jing-Chun Liau
Department of Electronic Engineering, Ming Chi University of Technology, Taipei, Taiwan, R.O.C.

Yang-Hsin Fan
Department of Computer Science and Information Engineering, National Taitung University, Taiwan, R.O.C.

Jan-Ou Wu
Department of Electronic Engineering, De Lin Institute of Technology, Taiwan, R.O.C.

Jhen-Ji Wang
Department of Electronic Engineering, Yuan Ze University, Chungli, Taoyuan, Taiwan, R.O.C.

ABSTRACT: In this paper, a new CMOS wide-band Low-Noise Amplifier (LNA) is proposed. It is operated in a range of 100 MHz–3.5GHz with current reuse, resistor feedback, mirror bias and inductor peaking technique. A two-stage topology is adopted to implement the LNA based on the TSMC 0.18-um RF CMOS process. Traditional resistor feedback wide-band LNAs suffer from a fundamental trade-off in gain, bandwidth and gain flatness. Therefore, we propose a new LNA which obtains high gain, wider bandwidth and good gain flatness performance by proposed technique. The simulation results show that the power gain is between 20.42 and 21.7dB (S21), the NF is less than 2.26dB, the input reflection coefficient (S11) is less than −10.62, and the third-order intercept point (IIP3) is about −14.1dBm. The LNA consumes maximum power at about 8 mW with a 1V power supply.

Applied System Innovation – Meen, Prior & Lam (Eds)
© 2016 Taylor & Francis Group, London, ISBN 978-1-138-02893-7

Delay-dependent stabilization condition for a class of T-S fuzzy systems with state and input delays

Shun-Hung Tsai, Yu-An Chen & Yu-Wen Chen
Graduate Institute of Automation Technology, National Taipei University of Technology, Taipei, Taiwan

ABSTRACT: The stabilization problem for a class of Takagi-Sugeno (T-S) fuzzy systems with state and input delays has been explored in this paper. First, the fuzzy controller has been designed via a Parallel Distributed Compensation (PDC) technique to stabilize the T-S fuzzy time-delay system. In addition, based on the Lyapunov-Krasoviskii function, the stabilization condition is formulated by the Linear Matrix Inequalities (LMIs). Furthermore, adopting the Right-Hand-Size (RHS) slack variable matrices technique, a less conservative stabilization condition for the T-S fuzzy time-delay system has to be obtained. Lastly, an example is illustrated to demonstrate the proposed stabilization condition that is less conservative and shows the validity of the proposed fuzzy controller.

Applied System Innovation – Meen, Prior & Lam (Eds)
© 2016 Taylor & Francis Group, London, ISBN 978-1-138-02893-7

An 8-bit 3.2MS/s low power SAR ADC for high-resolution audio applications

Chi-Hsiung Wang, Jen-Shiun Chiang, Wei-Ming Hsu, Wei-Bin Yang & Horng-Yuan Shih
Department of Electrical Engineering, Tamkang University, Taiwan, R.O.C.

Yu-Lung Lo
Department of Electronic Engineering, National Kaohsiung Normal University, Taiwan, R.O.C.

ABSTRACT: This paper presents a Successive Approximation Register Analog-to-Digital Converter (SAR ADC) design for High-Resolution applications. The SAR ADC used an energy-saving switching sequence technique for low power consumption. The simulation signal-to-noise-and-distortion ratio (SNDR) of the SAR ADC is 49.43 dB at 3.2 MS/s sampling rate. And the low power consumption is only 108.2-μW form a 1.8-V supply voltage. The SAR ADC is fabricated in a 0.18-μm CMOS technology.

Applied System Innovation – Meen, Prior & Lam (Eds)
© 2016 Taylor & Francis Group, London, ISBN 978-1-138-02893-7

A 200 MHz 23 mW high-efficiency Inductive Link Power Supply circuit with differential-driven CMOS rectifier and multiple LDOs in 0.18 μm CMOS process

Cheng-Wei Yang, Horng-Yuan Shih, Wei-Bin Yang & Chi-Hsiung Wang
Department of Electrical Engineering, Tamkang University, Tamsui, Taipei, Taiwan

ABSTRACT: A 200 MHz CMOS near-field Inductive Link Power Supply (ILPS) supplying output power of 18 mW and 5 mW under supply voltage of 1.8 V and 1 V is proposed in this paper. The CMOS power supply consists of differential-driven rectifier and Low-Dropout Regulators (LDOs). Two fully-integrated LDOs with the rectifier produce output voltage 1.8 V and 1 V for supplying analog and digital circuits, respectively. The rectifier has a cross-coupled bridge configuration and is driven by a differential RF input. The simulation output Power Conversion Efficiency (PCE) is 73.7% under 10 mA output current.

Applied System Innovation – Meen, Prior & Lam (Eds)
© 2016 Taylor & Francis Group, London, ISBN 978-1-138-02893-7

Implementation of an ultralow-voltage digitally controlled LDO in 0.18-μm CMOS technology

Y.-L. Lo, B.-Y. Liu & X.-H. Xiang
Department of Electronic Engineering, National Kaohsiung Normal University, Kaohsiung, Taiwan

W.-B. Yang
Department of Electrical Engineering, Tamkang University, New Taipei, Taiwan

ABSTRACT: This study presents an ultra-low supply voltage Low-Dropout regulator (LDO) with a digitally controlled technique. Based on a 0.18 μm standard CMOS process with $V_{TN} \approx 0.30$ V, $V_{TP} \approx 0.45$ V and VDD = 1.8 V. Unlike conventional analog LDOs, the proposed digitally controlled LDO does not require external output capacitor to stabilize the control loop. Furthermore, the proposed LDO uses a dual-loop architecture that contains a coarse-tuning loop for fast tracking and a fine-tuning loop for (process, voltage, and temperature) PVT variations tolerance. The experimental results demonstrate that the proposed LDO can operate from 0.7 V to 0.9 V with a dropout voltage of 200 mV. With a supply voltage of 0.9 V, the proposed LDO is capable of providing a regulated output of 0.7 V and delivering a maximal load current of 50 mA at 99.99% current efficiency.

Applied System Innovation – Meen, Prior & Lam (Eds)
© 2016 Taylor & Francis Group, London, ISBN 978-1-138-02893-7

A transient enhanced LDO with current buffer for SoC application

Wei-Bin Yang, Yu-Yao Lin, Ming-Hao Hong, Yin-Cheng Lin & Kuo-Ning Chang
Department of Electrical Engineering, Tamkang University, Tamsui, Taipei, Taiwan, R.O.C.

Yu-Lung Lo
Department of Electrical Engineering, National Kaohsiung Normal University, Kaohsiung, Taiwan, R.O.C.

ABSTRACT: This paper presents an analog transient enhanced method for LDO to reduce the over-/under-shoot voltage, and transient time in load transient response. When the output loading is changed, the current buffer provides an additional current path to accelerate the motion of power MOSFET. The LDO supplies 1.2 V to the System-on-a-Chip (SoC) core block, with its output voltage accuracy being less than 1.5% with no ESR compensation until 20 mA. The measurement result indicates that the quiescent current with band-gap reference is 40 uA. According to the experiment, the load transient time is less than 0.2 us when the output capacitance is 10 pF, and 15 us when the output capacitance is 1 uF. Moreover, the line transient response is less than 0.01% when the input voltage had 10% variation. Therefore, the LDO is suitable for fast transient, high accuracy SoC application.

Applied System Innovation – Meen, Prior & Lam (Eds)
© 2016 Taylor & Francis Group, London, ISBN 978-1-138-02893-7

Detection of short-circuit faults in induction machines through harmonics of the neighboring magnetic field—experimental and Finite Element investigations

V. Firețeanu & R. Pușcă
EPM_NM Laboratory, Politehnica University of Bucharest, Bucharest, Romania

ABSTRACT: The signature of a short-circuit fault in the stator winding of induction motors on the neighboring magnetic field is emphasized through a dedicated Finite Element (FE) model. A sensor system with two coils is proposed for the investigation of this magnetic field in order to detect the fault. The paper conclude on the possibility to detect incipient short-circuit faults in the stator winding through harmonics of the coil sensors voltage for no-load motor operation.

Applied System Innovation – Meen, Prior & Lam (Eds)
© 2016 Taylor & Francis Group, London, ISBN 978-1-138-02893-7

A voltage surveillance based switching controller with expandability

Chian-Yi Chao
Department of Electronic Engineering, Kao Yuan University, Taiwan

Chin-Ming Hsu
Department of Information Technology and Applications, Kao Yuan University, Taiwan

ABSTRACT: The main goal of this paper is to develop an expandable switching controller based on voltage surveillance, which could be applied to the sealed de-carbonized furnace to automatically control the ON/OFF of the switching valves and accurately adjust the quantities of the gas flow. The proposed expandable switching controller could support the workers/staffs and the company with the advantages of reducing the rates of the defective products, producing the optimal de-carbonized products precisely, and working in the safer circumstances and environment by timely controlling process temperature, thereby, reducing the explosion probability of the sealed furnace. In this paper, the proposed approach utilizes the microcontroller named MCS-51chip as the central control unit to sense the variation of the inputs, make the correct judgment on the gas flow control, output the warnings, and control the switching valves. The inputs of the proposed technology include the voltage values converted from the detected oxide, dioxide, monoxide, hydrogen, nitrogen, pressure, and temperature by using different infrared sensors, respectively. The outputs of the proposed switching controller mainly include the indicators and switching ON/OFF of gas flow control valves. After evaluating the experimental results, the proposed switching controller could support two distinctive contributions. (1) The proposed method of the microcontroller based technology can automatically and precisely control the ON/OFF of the switching valves by monitoring voltage variation. (2) This work could be extended and applied to the energy substitute switching control in order to supply the lasting power during the competition.

Applied System Innovation – Meen, Prior & Lam (Eds)
© 2016 Taylor & Francis Group, London, ISBN 978-1-138-02893-7

A high-efficiency interleaved DC-DC converter with zero-voltage-switching

Rou-Yong Duan & Jun-Yuan Lu
Department of Safety, Health and Environmental Engineering, Hung Kuang University, Taichung City, Taiwan, R.O.C.

ABSTRACT: The aim of this paper is to develop a step-up/step-down zero-voltage-switching interleaved DC-DC converter for the switching power supply. The interleaved converter is composed of two shunted elementary boost conversion units, including four switches, a three windings transformer and a small auxiliary inductor. Due to the inductor in series the input source, the duty cycle of two main switches can be controlled over 50% so that the voltage gain is higher than the turns ratio. The proposed converter, only applied a winding and a capacitor in series on secondary side, can be auto balanced the currents of the interleaved two phases without an additional current-sharing module. The leakage inductance problem of transformer is solved by the active clamped circuit that results the switches with the ZVS and synchronous property. The experimental results based on a 2 kW 155V to 500 V DC/DC prototypes and the maximum efficiency of the entire high-efficiency power conversion system is over 96.5% to verify the effectiveness of the theoretical analysis.

Applied System Innovation – Meen, Prior & Lam (Eds)
© 2016 Taylor & Francis Group, London, ISBN 978-1-138-02893-7

Improvement of multilevel inverters battery-balancing performance

Chia-Hsuan Wu & Liang-Rui Chen
Department of Electrical Engineering, National Changhua University of Education, Changhua, Taiwan

ABSTRACT: In this paper, an enhanced-battery-balancing strategy for seven-level inverter is proposed. There are three separate battery strings as the input voltage, the cascaded multilevel inverter and the digital controller. The output seven-level approximate sinusoidal waveform is produced by controlling the conduction period of different levels. The advantages of the proposed multilevel inverter are reducing the switching loss and the harmonic content as well as suitable for high-power applications. By proposed battery-balancing discharging strategy, two higher-capacity batteries are discharged to the load and also charge the lowest-capacity battery to prolong the available time of whole battery strings, efficiently use the battery energy and achieve the battery-balancing discharge. A multilevel inverter is designed for the system with the aged batteries to verify its feasibility and excellence. Finally, simulation and experimental results show that the proposed battery-balancing strategy for multilevel inverters actually and efficiently achieves the battery-balancing discharging function.

Applied System Innovation – Meen, Prior & Lam (Eds)
© 2016 Taylor & Francis Group, London, *ISBN 978-1-138-02893-7*

Study on voice signal control based LED lamp prototyping

Yu-Cherng Hung & Wan-Jung Lai
Department of Electronic Engineering, National Chin-Yi University of Technology, Taichung, Taiwan, R.O.C.

ABSTRACT: This paper studies a voice signal control based LED lamps. According to the frequency spectrum of the user voice, different LED lamps will be controlled so that they are either turned-on or turned-off. The architecture of the prototyping is composed of a pre-amplifier, bandpass filter, rectifier, integrator, comparator, and latch/toggle circuits. The system hardware is understood only by using conventional electronic devices without using the micro-processor single chip to enhance the stability and reduce the fabricated cost of the whole system.

Applied System Innovation – Meen, Prior & Lam (Eds)
© 2016 Taylor & Francis Group, London, ISBN 978-1-138-02893-7

Image contrast enhancement by Improving Histogram Equalization algorithm

Bing-Fei Wu & Meng-Liang Chung
Department of Electrical and Control Engineering, National Chiao Tung University, Hsinchu, Taiwan

Chih-Chung Ting & Chung-Cheng Chiu
Department of Electrical and Electronic Engineering, National Defense University, Taoyuan, Taiwan

ABSTRACT: Image enhancement techniques are widely applied in many fields. The purpose of image enhancement is to let the details of an image become clearer and to make the processed images more suitable for observers. Histogram Equalization (HE) is one of the well-known image enhancement methods because it is simple and effective. However, it is rarely applied to consumer electronics products because it may cause the excessive contrast enhancement and the feature loss problems that result in the processed images with unnatural look and unwanted visual artifacts. This study proposes a contrast enhancement algorithm, Improved Histogram Equalization (IHE), to solve the abovementioned drawbacks of HE. Furthermore, IHE enhances the weak features of images and makes the processed images have better visual quality. The experimental results show that images processed by IHE have better contrast and are more suitable for human visual perception than those of applying HE and other HE-based methods.

Applied System Innovation – Meen, Prior & Lam (Eds)
© 2016 Taylor & Francis Group, London, ISBN 978-1-138-02893-7

Controllable Magnetic Brake for wind turbines

Liang-Rui Chen, Yung-Shun Lin & Chia-Hsuan Wu
Department of Electrical Engineering, National Changhua University of Education, Changhua, Taiwan

Jonq-Chin Hwang
Department of Electrical Engineering, National Taiwan University of Science and Technology, Taipei, Taiwan

Chuan-Sheng Liu
Department of Aeronautical Engineering, National Formosa University, Yunlin, Taiwan

ABSTRACT: This paper presents the Controllable Magnetic Brake (CMB) to improve breaking performances of a conventional electrical brake for permanent magnet wind turbines. The proposed CMB can provide the extra torque to fix the blades not to rotate and resolve the disadvantage of exceeding current and power when the wind turbine works at abnormal wind velocities. The experiments show that the braking torque can be controlled and extended to 10 times as compared with a conventional electrical brake.

Applied System Innovation – Meen, Prior & Lam (Eds)
© 2016 Taylor & Francis Group, London, ISBN 978-1-138-02893-7

Extraction of the engineering inverse algorithm for LED laser eutectic die bonding process parameters

Wei-Yi Chan
Department of Mechanical and Electromechanical Engineering, National Sun Yat-Sen University, Kaohsiung, Taiwan

Chao-Ming Hsu
Department of Mechanical Engineering, National Kaohsiung University of Applied Science, Kaohsiung, Taiwan

ABSTRACT: This study proposes a 940 nm diode laser bonding process for LED chips. The key thermal parameters in the LED laser eutectic die bonding process are measured by employing inverse engineering analysis in this work. A Gaussian distributed laser was focused on the back side of an AlN substrate to provide the heat source in the bonding process. The eutectic Au80Sn20 solder metallized patched between the die and substrate is soldered by the heat conducted from the laser. The finite element software MSC. Marc was employed to simulate the bonding process. The temperature dependent material properties in coordination with the thermal-elastic-plastic FE model were used in simulation analysis. In comparing the simulated and measured temperature distribution patterns, which could maintain the numerical temperature results' difference within 10% as compared with that from the laboratory measurements.

Applied System Innovation – Meen, Prior & Lam (Eds)
© 2016 Taylor & Francis Group, London, ISBN 978-1-138-02893-7

On-line parameter estimation for position sensorless Permanent-Magnet Synchronous Generator drive

Y.C. Chang, W.F. Dai, B.H. Lien & S.Y. Wang
Department of Electrical Engineering and Advanced Institute of Manufacturing with High-Tech Innovations, National Chung Cheng University, Taiwan

ABSTRACT: This paper employs the on-line parameter estimation to improve the position sensorless performance of a Permanent-Magnet Synchronous Generator (PMSG) drive. The generator resistance will vary with the operating current and temperature, which causes the estimating error of the rotor angle. Therefore, the accurate resistance estimation of the PMSG will improve the position sensorless performance. This PMSG drive system consists of feedback circuits, the current control loop, position sensorless algorithm and on-line parameter estimation scheme. The dynamic model of a PMSG is established for the development of current control, position sensorless and on-line parameter estimation. The current control and extended-EMF position sensorless control are demonstrated. The recursive least squares method is implemented to estimate the resistance of the PMSG. The implementation of PMSG resistance estimation will improve the accuracy of the rotor position estimation. All the control and estimation algorithms are digitally fulfilled by a microcontroller. The experimental results verify the accuracy of the estimated parameter and the improvement of the position sensorless algorithm.

Applied System Innovation – Meen, Prior & Lam (Eds)
© 2016 Taylor & Francis Group, London, ISBN 978-1-138-02893-7

Smart control module design for IoT applications by a power meter SoC

Hsing-Feng Chen, You-Ting Lin, Cheng-Hong Wu, Zheng-Han He,
Bo-Lin Liou & Yi-Kai Liao
Department of Electrical Engineering, Cheng Shiu University, Kaohsiung, Taiwan

ABSTRACT: This manuscript provides a smart control module design, used for IoT applications, by a single phase power meter SoC. This design integrates smart functions into a compact controller to easily control and monitor the electrical appliances. The smart functions including communication, time scheduling, power monitoring, and safety protections et al., can be selectable and adjustable, depending on the objects which are controlled or monitored. In this module, wired or wireless communication can be performed for the data gather or control of appliance. Even the parameters of environment where appliances installed can be also monitored. Embedded with the power meter functions, this module can record the electricity data, including current, voltage, power, and even electricity cost of appliance. Time scheduling to turn on/off device is available and can be easily regulated. The safety alarms such as leakage current detection, over-current and unbalanced phase, are also built-in. Thus, with these functions, this module can be effectively used to implement an IoT node and to build the IoT network easily.

Applied System Innovation – Meen, Prior & Lam (Eds)
© 2016 Taylor & Francis Group, London, *ISBN 978-1-138-02893-7*

Blend novel recurrent modified Gegenbauer OPMPSONN control for a six-phase copper rotor IM servo-drive CVT system

Chih-Hong Lin
Department of Electrical Engineering, National United University, Miaoli, Taiwan

ABSTRACT: In order to reduce efforts from uncertainties in the Continuously Variable Transmission (CVT) system driven by a six-phase copper rotor Induction Motor (IM), a blend novel recurrent modified Gegenbauer Orthogonal Polynomial Modified Particle Swarm Optimization Neural Network (OPMPSONN) control system is proposed to minimize the influence from uncertainties in this paper. As the blend novel recurrent modified Gegenbauer OPMPSONN control system can learn uncertainties with nonlinear and time-varying characteristics, the proposed control system has a much better control performance than the linear controller in the occurrence of uncertainties. To obtain better and faster convergence, the modified PSO method is used to adjust two learning rates of parameters in the recurrent modified Gegenbauer NN. At last, the effectiveness of the proposed control system is verified by experimental results through the demonstration of comparative studies.

Applied System Innovation – Meen, Prior & Lam (Eds)
© 2016 Taylor & Francis Group, London, *ISBN 978-1-138-02893-7*

Fast Maximum Power Point Tracking for Photovoltaic Generations

Jen-Hao Teng, Wei-Hao Huang, Tao-An Hsu & Chih-Yen Wang
Department of Electrical Engineering, National Sun Yat-Sen University, Kaohsiung, Taiwan

ABSTRACT: A fast Maximum Power Point Tracking (MPPT) for Photovoltaic Generations (PVGs) is proposed in this paper. Based on the measured voltage and current of a PVG, the proposed MPPT uses the characteristic output function of the PVG and Newton-Raphson method to calculate the solar irradiance and temperature. The maximum power point can then be found directly. Since only two measurement sets are enough for the proposed MPPT to calculate the maximum power point, the proposed MPPT can effectively reduce the transient response time of the conventional Perturbation and Observation (P&O) method and increase the harvested energy during MPPT. Simulations by MATLAB for the proposed MPPT and P&O MPPT are used to verify the feasibility of the proposed MPPT. Experimental results show that the shorter transient response time and larger energy during MPPT can be achieved by the proposed MPPT.

Applied System Innovation – Meen, Prior & Lam (Eds)
© 2016 Taylor & Francis Group, London, ISBN 978-1-138-02893-7

Automatic detection design and electromagnetic shielding effectiveness of synthetic graphite sheets

Rong-Jong Wai
Department of Electronic and Computer Engineering, National Taiwan University of Science and Technology, Taipei, Taiwan

Yeou-Fu Lin
Department of Electrical Engineering, Yuan Ze University, Chung Li, Taiwan

Chiung-Chou Liao
Department of Electronic Engineering, Chien Hsin University of Science and Technology, Chung Li, Taiwan

ABSTRACT: The electronics industry has made great strides in a short time, with a trend of slimness and lightness which are now ubiquitous in laptops and tablets. When the conductivity of copper or aluminum did not seem satisfactory enough, synthetic graphite sheets made an appearance with a higher heat transfer coefficient (1350~2100 W/mk). Synthetic graphite and its derived composites were created as a result of these demands. At present, Taiwan's synthetic graphite sheets are currently imported. Not only are they expensive, but also it is impossible to verify their quality. In the past, it was used to cut a large synthetic graphite sheet into several small ones to be detected. This study consists of two sections: the first deals with research and development of an automatic detection platform with thin and lightweight materials as primary composites, which enables continuity check for production purpose without the need to conduct destructive tests. This study has achieved a repetition rate of 1.05458% which is better than 2.38964% obtained using Laser Flash Method. Its high precision enables subsequent grade classification. This study mainly aims to considerably reduce workforce as well as the risk associated with operations in high temperatures, and increase the degree of automation. The second section does research into validation of electromagnetic shielding effectiveness of synthetic graphite sheets, which is measured in the frequency bands including 300 MHz, 900 MHz, 1800 MHz, and 2450 MHz. These four bands are frequently-used microwave and RF bands. The results indicate that a high shielding effectiveness of 99.99% can be achieved in these four frequency bands.

Applied System Innovation – Meen, Prior & Lam (Eds)
© 2016 Taylor & Francis Group, London, ISBN 978-1-138-02893-7

Study on the effect of temperature on CMOS transistor during chip designing

Y.-C. Hung & Y.-H. Hsieh
Department of Electronic Engineering, National Chin-Yi University of Technology, Taichung, Taiwan, R.O.C.

ABSTRACT: This research studies the effect of temperature on MOS transistors. The characteristics of NMOS and PMOS transistors with different sizes, operating at a range of –40°C to 80°C, was investigated using the HSPICE simulation. HSPICE is a powerful circuit simulator, especially in chip designs. The current of MOS transistor greatly varies with temperature variations. Two major factors, threshold voltage variation and mobility degradation, are studied in this work. After inspecting the simulation results, we find that the increased or decreased current of MOSFET will depend on the transistor's operation region and transistor size. In addition, some important electrical analog parameters, such as gain, trans-conductance gm, output conductance gds, and the frequency variation of Voltage-Control-Oscillator (VCO), are also investigated. CMOS 0.18-um and 0.35-um process technologies are used to enhance generality and effectiveness.

Mechanical & Automation Engineering

Applied System Innovation – Meen, Prior & Lam (Eds)
© 2016 Taylor & Francis Group, London, ISBN 978-1-138-02893-7

The comprehensive dynamics simulation of the gearbox housing based on Patran

Yi-Ming Lee
Department of Mechanical and Automation Engineering, Kao Yuan University, Kaohsiung, Taiwan

Chao-Wu Liu
Der Young Technology Co., New Taipei, Taiwan

ABSTRACT: The process of product design in the industry has been significantly refined over the years through the capabilities of advanced computer aided design and engineering tools. Conventional techniques, based on calculations, for implementing the simulation of a gearbox are time consuming.. This paper illustrates the significant benefits in implementing the efficiency of a gear reducer, by using a correlation solid dynamics simulation methodology design available in the MSC/Patran. A case study will be provided for improving the correlation potential both in the stress and strain. The main motivation behind the work is to go for a complete FEA of the gearbox housing rather than the empirical formulae and iterative procedures. In order to describe the better process of the gearbox simulation, 3D, CAD, and ADAMS are discussed and verified to construct the elemental models for a specific gear box in the comprehensive design.

Applied System Innovation – Meen, Prior & Lam (Eds)
© 2016 Taylor & Francis Group, London, ISBN 978-1-138-02893-7

Experimental measurement of thermal strain using optical fiber sensors

Shiuh-Chuan Her & Chih-Ying Huang
Department of Mechanical Engineering, Yuan Ze University, Chung-Li, Taiwan

ABSTRACT: An analytical solution is presented to predict the thermal stress and strain of a surface bonded optical fiber sensor induced by a temperature change. It shows that the thermal strain transferred from the host structure to the optical fiber is strongly dependent on the bonding characteristics among the optical fiber, protective coating, adhesive layer and the host material. Experimental tests were conducted using fiber Bragg grating sensor to measure the thermal strain of the surface bonded optical fiber. Good agreements were observed in comparison with the experimental results and theoretical calculations.

Applied System Innovation – Meen, Prior & Lam (Eds)
© 2016 Taylor & Francis Group, London, ISBN 978-1-138-02893-7

Spherical object recognition via a novel fuzzy-based color correction method

Shun-Hung Tsai, Yu-Wen Chen & Yu-An Chen
Graduate Institute of Automation Technology, National Taipei University of Technology, Taipei, Taiwan

ABSTRACT: In the real world, color variations in spherical objects are not identical under high brightness conditions, and it will make the recognition difficult or result in a misrecognition. In order to mitigate this problem, this study proposes a fuzzy-based color correction method for spherical object image recognition. By adopting a fuzzy logic scheme, the hue angle for color distortion areas can be determined and the color distortion area can also be recognized accurately. In addition, a design procedure for image recognition of spherical objects under high brightness conditions is proposed in this paper. Some examples are illustrated to show the feasibility and effectiveness of the proposed color correction method for spherical object recognition.

Applied System Innovation – Meen, Prior & Lam (Eds)
© 2016 Taylor & Francis Group, London, ISBN 978-1-138-02893-7

A constrained Model Predictive Control for Dual-Stage Actuator systems with improved tracking performance

Chi-Ying Lin & Chih-Yuan Chang
Department of Mechanical Engineering, National Taiwan University of Science and Technology, Taipei, Taiwan

ABSTRACT: This paper presents a constrained Model Predictive Control (MPC) algorithm which aims to improve the tracking control performance of practical dual-stage actuator systems. To efficiently facilitate the management of control action in each actuator for better transient performance, a new objective function by adding an extra error weighting on a selected actuator is applied to obtain optimal control moves during the on-line optimization process. The dual-stage actuator system adopted for tracking performance evaluation includes a servo motor driven linear stage as the coarse actuator and a stacked piezoelectric actuator as the fine actuator. To reduce the computational burden and avoid the possibly-deteriorated performance caused by the use of conventional hard constraints, an efficient constraint condition which particularly handles the fine actuator travel limit is derived for real implementation because fast sampling is essential for the servo design of piezoelectric actuator. Comparative experiments on tracking set-point reference profiles demonstrate the effectiveness of the proposed constrained MPC algorithm.

Applied System Innovation – Meen, Prior & Lam (Eds)
© 2016 Taylor & Francis Group, London, ISBN 978-1-138-02893-7

Machinability analysis and setup optimization for five-axis machining

Wei-chen Lee & Ching-chih Wei
Department of Mechanical Engineering, National Taiwan University of Science and Technology, Taipei, Taiwan

ABSTRACT: Two potential problems that are usually ignored by the CAM software for five-axis machining: one is the overtravel for the three linear axes of the 5-axis machine tool; the other is the collision between the spindle and the worktable. The objective of this research was to propose a method based on the visibility map and then develop a computer program to prevent the mentioned problems from happening. In addition, the optimal position for the workpiece and the optimal tool axis along the tool paths can subsequently be determined through the proposed method. Two case studies were performed to demonstrate that the method proposed in the paper is feasible.

Applied System Innovation – Meen, Prior & Lam (Eds)
© 2016 Taylor & Francis Group, London, ISBN 978-1-138-02893-7

Task allocation and path planning for service robots based on swarm intelligence algorithms and Wireless Sensor Network localization

Kuan-Yu Chen, Chien-Hung Chen, Wei-Ching Wang & Ho-Kai Wang
Department of Mechanical Engineering, Chung Yuan Christian University, Chungli, Taoyuan, Taiwan

ABSTRACT: This paper proposes an integrated approach of task allocation, path planning and navigation for service robots based on swarm intelligence algorithms and wireless sensor network localization. Assume that two service robots serve in an intelligent office building for delivering documents or packages, and all staff in different locations can assign tasks to the robots. Therefore, the central management system in this workplace must have dynamic task allocation and path planning abilities for control of service robots. In this paper, we firstly perform task allocation for service robots by using particle swarm optimization, then a modified ant colony optimization is applied to plan the shortest path for each robot under environmental constrains in the workplace. Namely, we combine the two swarm intelligence algorithms mentioned above to perform dynamic task allocation and path planning for the service robots. In addition, an indoor positioning system and its error compensation for service robots are developed using ZigBee wireless sensor networks and a stereo vision computational model, respectively. We also integrate a number of interactive web technologies to develop a web application with friendly graphical user interface for users who can assign tasks and monitor real-time statuses of service robots. Finally, experimental results are given in this paper to demonstrate the effectiveness of the integrated proposed approach which allows users to assign tasks on the Internet and then each service robot can complete the assigned tasks along a feasible path.

Applied System Innovation – Meen, Prior & Lam (Eds)
© 2016 Taylor & Francis Group, London, ISBN 978-1-138-02893-7

Study on the characteristics of oil-magnet hybrid hydrodynamic bearing at low speed

Yi-Hua Fan, Liao-Yong Lou & Hsiang-Jung Chang
Department of Mechanical Engineering, Chung Yuan Christian University, Taoyuan City, Taiwan, R.O.C.

ABSTRACT: Fluid hydrodynamic bearings are not suitable to operate at low speed and will be damaged as the rotor starts to operate. This study proposes a passive magnetic bearing combined with the fluid hydrodynamic bearings to improve the drawbacks and make the magnet-fluid hybrid hydrodynamic bearings maintain better characteristics at low speed. It can be proved that the magnetic force of permanent magnets can support the suspension of the spindle in the hydrodynamic bearing with little gaps. These gaps will make the rotor and the stator not to contact each other as they stop or at a low rotating speed. Thus, there is no dry friction in the bearing and the lifetime of the bearing will be increased.

Applied System Innovation – Meen, Prior & Lam (Eds)
© 2016 Taylor & Francis Group, London, ISBN 978-1-138-02893-7

Time-delayed dynamic response and control of Shape Memory Alloy composite beam subjected to random excitation

X.M. Li, Z.W. Zhu & J. Xu
Department of Mechanics, Tianjin University, Tianjin, China

ABSTRACT: The time-delayed dynamic responses of Shape Memory Alloy (SMA) composite beam subjected to stochastic excitation are investigated. Nonlinear differential items are introduced to explain the hysteretic phenomena of SMA, and the nonlinear time-delayed dynamic model of SMA composite beam subjected to stochastic excitation is developed. The stochastic stability of the system is analyzed, and the probability density function of the system's response is obtained. Finally, the conditions of stochastic Hopf bifurcation are determined. Numerical simulation results show that the stability of the system varies with bifurcation parameters, and stochastic Hopf bifurcation appears in the process. The results obtained in this research are helpful for engineering applications of SMA composite beam.

Applied System Innovation – Meen, Prior & Lam (Eds)
© 2016 Taylor & Francis Group, London, ISBN 978-1-138-02893-7

Study of automatic baud rate detection technology

Wen-Cheng Pu & Hung-Che Chiung
Department of Electrical Engineering, National Chin-Yi University of Technology, Taichung, Taiwan

Meei-Ling Hung
Department of Electrical Engineering, Far-East University, Tainan, Taiwan

ABSTRACT: An RS232 interface, a low-speed serial communication protocol, is widely applied to the general and the industrial communication, but the communication between the devices is executed under the limitation of the setting action of the baud rate and the frame format parameters. However, the parameter setting manner becomes an important problem for a long-distance communication. The development of the automatic baud rate setting technology intends to solve this problem by adding extra hardware devices or editing extra firmware. These methods may take more time, increase the cost of development and cause the instability of the system. This study, capable of being compatible with the hardware and the firmware of the conventional interface, can combine with the PKII encoding technology to build a command channel that need not set parameters and possesses the automatic communication bandwidth adaptation between the traditional CTS/RTC pins, thereby matching the parameter settings between two ends of the communication automatically, promoting the accuracy of the handshaking process and providing more functions. This study verifies its viability at the last stage by experimenting with a personal computer in combination with the Arduino UNO board.

Applied System Innovation – Meen, Prior & Lam (Eds)
© 2016 Taylor & Francis Group, London, ISBN 978-1-138-02893-7

Study of yield stress effect on a new Magnetorheological brake

Y. Shiao & N.A. Ngoc
Department of Vehicle Engineering, National Taipei University of Technology, Taipei, Taiwan

ABSTRACT: Magnetorheological (MR) brake provides fast response and good controllability for many low-torque applications. The maximum brake torques for most of MR brakes are still not high enough because the volume and mass of an MR brake are still large. The maximum brake torque of an MR brake is decided by the generated magnetic field and consequent yield stress. To reduce the ratio of torque to volume (or torque to mass), yield stress of an MR brake must be explored and raised by some effective methods. We propose a new design of MR brake, which features multiple magnetic poles and multiple MR fluid layers. Design of single electro-magnetic pole is very popular in most of published MR brakes. A 4-pole 2-layer MR brake is proposed in this paper to increase the maximum brake torque. Beside, the yield stress of this brake is analyzed and improved. A 3-D magnetic simulation is performed for the MR brake, and the yield stress of MR brake is observed carefully to check the influences of brake diameter, ampere turn of coils, and ring thickness of rotor. Simulation results show that the yield stress increases linearly by increasing thickness of rotors. However, torque saturation occurs for the increase of a number of ampere turns of coils. Compared with the performance of a conventional MR brake with the same volume dimension, the maximum brake torque of this new brake is nearly two times larger. Finally, simulation results confirm the feasibility of this new brake of the maximum yield stress and brake torque enhancement.

Keywords: MR brake, yield stress, multiple pole

Applied System Innovation – Meen, Prior & Lam (Eds)
© 2016 Taylor & Francis Group, London, ISBN 978-1-138-02893-7

Wind-shear encountered landing control based on Adaptive FCMAC

J.G. Juang & T.C. Yang
Department of Communications, Navigation and Control Engineering,
National Taiwan Ocean University, Keelung, Taiwan

ABSTRACT: Intelligent aircraft Automatic Landing Systems (ALS) that use adaptive Fuzzy Cerebellar Model Articulation Controller (FCMAC) is being presented, in this paper. The control scheme adopts type-1 and type-2 FCMAC in the controller design. Stability theory is applied to control system analysis and adaptive learning rule derivatives to construct the Adaptive FCMAC (AFCMAC). The performance on tracking the desired landing path and environment adaptive capability are demonstrated through software simulations. The proposed controllers can guide the aircraft to a safe landing in severe wind shear conditions.

Applied System Innovation – Meen, Prior & Lam (Eds)
© 2016 Taylor & Francis Group, London, ISBN 978-1-138-02893-7

Gear-shift control strategy for automatic transmission based on clutches' pressure

Q.K. Wei, Y.L. Lei, B.Q. Hu & Z.W. Liu
National Key Laboratory for Automotive Simulation and Control, Changchun, China

X.Z. Li
Hangzhou Advance Gearbox Group Co. Ltd., Hangzhou, China

ABSTRACT: The gear-shift process of automatic transmissions usually needs an engaged and disengaged clutch, which is called the clutch-to-clutch shift control. One of the key factors affecting the shift quality is the pressure control of the engaged and disengaged clutches. First, the dynamic analysis of the gear-shift process for automatic transmission was studied based on this hydraulic control system. Second, an ideal pressure control curve was determined for the power-on upshift process of the on-coming and off-going clutches. Third, this paper proposed the pressure control strategy of shifting clutch based on feedback current of the corresponding solenoid valve and formed a feedforward closed-loop PID control system that had advantages in both the stability of close-loop control and the rapidity of feedforward control. Finally, we performed experimental verification on the vehicle, and the test result proved that the proposed clutch control strategy was proper and feasible.

Applied System Innovation – Meen, Prior & Lam (Eds)
© 2016 Taylor & Francis Group, London, ISBN 978-1-138-02893-7

Front impact analysis and design improvement for an electric all-terrain vehicle

J.S. Chen, H.Y. Hwang & Y.S. Chen
Department of Vehicle Engineering, National Taipei University of Technology, Taipei, Taiwan

ABSTRACT: Environmental friendly designs are one of the major focuses recently. Many automobile manufacturers have been continuing designing and producing environmental friendly vehicles. An existing combustion engine all-terrain vehicle was modified and converted to an electric street vehicle. The goal of this study is to analyze this electric vehicle, to develop a lightweight design improving its cruising abilities and to retrofit the front structure to enhance the frontal crashworthiness. Structure modification is needed due to different design requirements between two distinct vehicle functions, one for all-terrain rough road and the other for street drive. With the limited power provided by the given batteries, light weight design is critical. A design is proposed with a new bumper and a modified structure to absorb the impact energy. The proposed structure improved the head injury criteria by 35%, increased the residual space by 59%, and raised the energy absorption by 27%.

Applied System Innovation – Meen, Prior & Lam (Eds)
© 2016 Taylor & Francis Group, London, ISBN 978-1-138-02893-7

An experimental investigation on the effect of engine performance and exhaust emissions by using a by-pass cooling air compressor device in the internal combustion

Ming-Hsien Hsueh & Da-Fu Lin
National Kaohsiung University of Applied Sciences, Kaohsiung, Taiwan

ABSTRACT: This paper presents a new device for the internal combustion to increase the combustion efficiency and improve the exhaust gas emissions. An additional cooled air is input to the inlet manifold by the device to provide more comburent or combustibles for the engine. The device is set between the intake valve and the throttle. A reinforced air fan is designed in the device to input the additional air to the inlet manifold from the side of the manifold. The additional air not only increases the capacity of comburent or combustibles, but also speeds up the flow velocity of the intake air that is similar to the air multiplier technology. Because there are no blades on the inlet manifold, the intake airflow can be accelerated smoothly instead of the problem of the turbo lag in a turbo-charger engine. The Thermoelectric module Chip (TEC) is applied in the device to cool the additional intake airflow that can increase the oxygen density for combustion of the engine by the physical property of thermal expansion and contraction.

Applied System Innovation – Meen, Prior & Lam (Eds)
© 2016 Taylor & Francis Group, London, ISBN 978-1-138-02893-7

Finite element model updating of ventilated disc-brake rotors

F.T. Kao & K.N. Chen
Department of Mechanical Engineering, Tungnan University, New Taipei, Taiwan

W.H. Gau & C.Y. Hung
Department of Mechatronic Engineering, Huafan University, New Taipei, Taiwan

ABSTRACT: In most finite element studies, the finite element models are not verified but assumed to be correct. This assumption should be substantiated before conducting further finite element simulations. To achieve this purpose, the FE model updating technique can be adopted to correlate the FE analysis result to the experiment. In this research, experimental modal testing and finite element analysis on a ventilated disc-brake rotor were first performed, and then based on the test frequencies and mode shapes, FE model updating was carried out using geometric and material parameters as the updating parameters. Two cases exemplified that the FEA frequencies of the updated models were in excellent agreement with the experimental data. The resulted FE models are capable of accurately predicting the dynamic behaviors of the disc-brake rotor and suitable for further design modification studies.

Applied System Innovation – Meen, Prior & Lam (Eds)
© 2016 Taylor & Francis Group, London, ISBN 978-1-138-02893-7

Study on the application of magnetic force to mold decoration formation

Meng-Hsun Tsai & Shi-Chang Tseng
Department of Mechanical Engineering, National Yunlin University of Science and Technology, Yunlin, Taiwan, R.O.C.

ABSTRACT: This study is an innovative in-mold decoration (IMD) process. A peel test was performed on a pre-heated polymer thin film laminated with a piece of plastic using magnetic assisted molding to verify the feasibility of this type of molding in the IMD process, and to discover the optimal processing parameters such as the pre-heated temperature, holding time, and magnetized steel ball diameter.

Applied System Innovation – Meen, Prior & Lam (Eds)
© 2016 Taylor & Francis Group, London, *ISBN 978-1-138-02893-7*

Coupled bending and torsional out-of-plane vibrations model of a frame structure with partially clamped joints

Shueei-Muh Lin, Kun-Wei Lee & Wen-Rong Wang
Department of Mechanical Engineering, Kun Shan University, Tainan, Taiwan, R.O.C.

ABSTRACT: In practice, the joint of a frame may be partially or completely clamped. So far, its effect on the dynamic behavior has not been investigated clearly. In this study, these joints are mathematically simulated and generalized as an elastic joint associated with a rotational spring constant. The elastic joint is associated with the rotational angle difference of a joint between two connected elements. In addition, there exists the moment of resistance against the rotational angle difference. This moment of resistance is defined as a rotational spring constant. If the spring constant is infinite, the joint is completely clamped and the rotational angle difference is zero. On the other hand, if the spring constant is zero, the joint is hinged. Moreover, the bending vibration model of a general asymmetry frame with two elastic joints was established in this study. Its exact solution for the general system was derived. It was found that the effects of the rotational spring constants, the ratios of the bending rigidities, and the element lengths on the vibration of a frame were significant.

Keywords: out-of-plane vibrations; asymmetry frame structure; partially clamped joints

Green Technology & Architecture Engineering

Applied System Innovation – Meen, Prior & Lam (Eds)
© 2016 Taylor & Francis Group, London, ISBN 978-1-138-02893-7

Analysis of the offshore wind turbine structure with the earthquake influence

K.Y. Huang
Department of Bio-Industrial Mechanics Engineering, National Chung Hsing University, Taichung, Taiwan

G.C. Tsai & B.J. Tsai
Department of Mechanical and Electro-Mechanical Engineering, National I-Lan University, Yilan, Taiwan

ABSTRACT: This research mainly uses the finite element software to analyze the stress, strain, and deformation of the offshore wind turbine with the earthquake influence. The research results show that the maximum stress has exceeded the material's ultimate stress, which often occurs on the power system, and that it is important to pay attention to the design and construction of wind turbines.

Applied System Innovation – Meen, Prior & Lam (Eds)
© 2016 Taylor & Francis Group, London, ISBN 978-1-138-02893-7

Design and altitude control of a Kuroshio Generator System

Hsing-Cheng Yu, Guan-Yi Wu & Young-Zehr Kehr
Department of Systems Engineering and Naval Architecture, National Taiwan Ocean University, Keelung, Taiwan

ABSTRACT: In this study a Kuroshio Generator System (KGS) was designed to apply to the Keelung Sill, which is near the Taiwan maritime environment. The KGS combined a reliable cable and anchor subsystem and an adjustable flap that could smoothly operate at sea, and could ignore the motion changes of the rotation axes in yaw and roll via a proper rudder design. An intuitive simulation method was adopted in the MapleSim software to create a rigid body structure modeling of the KGS. The KGS stable requirements of the dynamic equilibrium, motion behavior, and the reliable cable included a combination of Y-shaped elastic polyester ropes; a rigid chain was analyzed and achieved. In addition, the KGS has advantages of altitude control in routine inspection or avoidance affected by the severe weather (e.g., typhoons or storms). The flap lift in the KGS could be adjusted by flap angles for altitude control. An optimal proportional-integral-derivative control law with Ziegler-Nichols tuning method was applied in the KGS to control the altitudes from 22.3 m to 16.5 m, and the rise time, settling time, and maximum overshoot of time response characteristics are 3.3 s, 9.7 s, and 3.3%, respectively. Furthermore, the estimated output power of the dual generators in the KGS has approached to 110 KW.

Applied System Innovation – Meen, Prior & Lam (Eds)
© 2016 Taylor & Francis Group, London, ISBN 978-1-138-02893-7

Measurements for tray fermented biological agents of fungus in a temperature–humidity controlled ambience

Rong-Yuan Jou & Jing-Jhong Cao
Department of Mechanical Design Engineering, National Formosa University, Yunlin/Huwei, Taiwan

ABSTRACT: Microorganism cultivation through Solid-State Fermentation (SSF) is a fermentation process on solid substrates in the absence or near absence of free water in the medium. Quantitative description about the influence of temperature and moisture on microbial growth and drying operations is essential for the modeling and optimization of SSF. Furthermore, drying of biological materials is subjected to the characteristics, such as a low drying temperature and a long drying time. This study aimed to measure the growth and the drying conditions of biological agents of Trichoderma fungus inside a precisely controlled temperature–humidity chamber. Moisture variations after fermentation and drying of different substrate layers were measured by MHE-800 and MHK-120 thermo-hygrostat apparatus, respectively. The variation history of the moisture ratio during different-thickness substrate drying by MHE-800 and by MHK-120 apparatus was investigated. Fermentation experiments were conducted at an ambient temperature of 28°C, a relative humidity of 98 RH% under 144 hours continuous fermentation with no mixing operations. Drying experiments were conducted at an ambient temperature of 40 °C and a relative humidity of 20 RH%, and the mixing operations were done for every hour until the substrate's moisture content reached the set point of 10% level. Experimental results of MHE-800 and MHK-120 systems were compared to explore the influences of designed system performances upon processing the results.

Applied System Innovation – Meen, Prior & Lam (Eds)
© 2016 Taylor & Francis Group, London, ISBN 978-1-138-02893-7

Development of web-based Intelligent Living System based on ZigBee technology

Yun-Wu Wu, Chin-Min Chen & Chao-Sen Yen
Department of Architecture, China University of Technology, Taipei, Taiwan

ABSTRACT: In recent years, owing to the rapid development of wireless sensor networks and mobile computing advances in technology based on the Internet of Things and smart phones with the support of ZigBee and sensor technology has been applied extensively, and ubiquitous control has come true. This study presents ZigBee technology into a proposed ubiquitous intelligent living system, with the use of the routing ZigBee gateway and external network, making all sorts of living appliances to connect to a web-based intelligent system based on wireless technology. With the usage of wireless technology, this proposed system can reduce the difficulties of the connections between the living appliances, and it is convenient for users to access or remove these appliances freely under a wireless or ubiquitous environment that is helpful to improve our life's comfort level.

Keywords: internet of things; ZigBee technology; Intelligent Living System; wireless

Applied System Innovation – Meen, Prior & Lam (Eds)
*© 2016 Taylor & Francis Group, London,**ISBN 978-1-138-02893-7*

Ecodesign: An example of take-away packaging

Hui-Ting Tang & Yuh-Ming Lee
Institute of Natural Resources Management, National Taipei University, New Taipei City, Taiwan

ABSTRACT: As global resource consumption has been increasing dramatically, our priority would be improving the rate of resource productivity. In design for the environment, products are thought in an ecological way from the moment they are conceived. Such a thinking echoes the product lifecycle management, in which any problems are better tackled at the design phase. This study aimed to provide a greener and an economically-feasible solution for take-away food packaging. The succinct design uses only a minimum number of natural resources and yet achieves the maximum functionality and sustainability, ensuring the well-being of both humans and the Earth.

Applied System Innovation – Meen, Prior & Lam (Eds)
© 2016 Taylor & Francis Group, London, ISBN 978-1-138-02893-7

Research on the control of the evaluation factors of the TVOC emissions from green building materials in Taiwan

Cheng-Chen Chen
Department of Interior Design, Tung Fang Design Institute, Taiwan, R.O.C.

Ching-Chang Lee
Department of Environmental Occupational Health, National Cheng Kung University, Taiwan, R.O.C.

Jui-Ling Chen
Architecture and Building Research Institute, Ministry of Interior, Taiwan, R.O.C.

ABSTRACT: In the subtropical climate of Taiwan, indoor building materials have a significant impact on the poor indoor air quality, which leads to health problems and increased risks of cancer. Currently, the "Green building materials labels" and the Indoor Air Quality Act have been promoted to control the source of pollution, namely the fugitive pollutants of the building materials and the air concentration in the indoor space. Emphasis is placed on the advanced administrative control of the Total Volatile Organic Compounds (TVOC) to reduce the harm that could be caused by the fugitive pollutants of the indoor building materials to health. This study adopted the fugitive emission database of 645 pieces of the building materials with a green label, test, and verification of the fugitive emission of the seven small pieces of the building materials, numerical statistics and analysis, and literature analysis to research the control of the evaluation factors of fugitive TVOC from green building materials in Taiwan. The purpose of this is to prevent other toxic Volatile Organic Compounds (VOCs) from affecting the indoor air quality, effectively controlling the harm to residents, reducing the energy consumption of the air conditioning for removing pollutants through ventilation, and reducing the carbon emissions, and protect the environment.

Applied System Innovation – Meen, Prior & Lam (Eds)
© 2016 Taylor & Francis Group, London, ISBN 978-1-138-02893-7

A study on energy performance of homestay industry in Taiwan

Ming-Yu Yang, Tain-Fung Wu & Shieh-Liang Chen
Department of Business Management, Asia University, Taichung, Taiwan

Hui-Yen Liao
Architecture and Building Research Institute, Ministry of Interior, Taipei, Taiwan

Shin-Ku Lee
Research Center for Energy Technology and Strategy, National Cheng-Kung University, Tainan, Taiwan

ABSTRACT: Tourism has emerged as one of the leading service industries that spurs economic development, providing the outlines of a better way of life for the public, promoting people-to-people diplomacy, and bringing many business opportunities. In Taiwan, the homestay sector has grown rapidly along with the trends of recreational farming and ecotourism. Since the homestay sector is booming in Taiwan, the related energy demands are also rising. To achieve the goal of green homestays, the energy consumption data and other pertinent information for 2013 is obtained from 29 approved homestays. Finally, the annual energy performance of individual homestays is evaluated in terms of EUI, which is defined as the source energy consumption per unit of gross floor area. The survey data indicated that the average EUI of homestays is 176.8 kWh/m^2. Electricity consumption accounts for 84.6% of total energy usage, Diesel oil consumption for hot-water boiler accounts for 15.2% of total energy usage. The averaged EUI of homestays obtained in this study is more than the EUI proposed by Wang's study in 2010 (144 kwh/m^2). This result implies that the growing yearly occupancy rate of homestay sector in Taiwan lead to the increment of EUI. Average EUI is not a fair assessment of the energy performance of homestay. To develop a benchmarking system of homestays in Taiwan is suggested in this study.

Keywords: homestay; energy usage intensity; benchmarking; green tourism

Applied System Innovation – Meen, Prior & Lam (Eds)
© 2016 Taylor & Francis Group, London, ISBN 978-1-138-02893-7

A thermal comfort evaluation of the container house using the thermal admittance simulation process

H.Y. Shih & Y.T. Chou
Department of Applied Geoinformatics, Chia Nan University of Pharmacy and Science, Tainan, Taiwan

S.Y. Hsia
Department of Mechanical and Automation Engineering, Kao Yuan University, Kaohsiung, Taiwan

B.W. Lee
Department of Tourism Management, Shih Chien University, Kaohsiung, Taiwan

ABSTRACT: In recent years, building energy efficiency is increasingly being valued. For architectural design, we not only consider its functionality but also evaluate the comfort requirement of its interior and exterior ventilation, environment control, and the use of the material. In this paper, commercial software, Autodesk Ecotect Analysis, is implemented to evaluate the thermal distribution of the container house for predicting the energy use and efficiency. In order to verify the feasibility of the simulation process, we compare the measured data and simulating results of a 20-feet long container house at Tainan City. Less than 10% error is considered as a good agreement and holds a promise for further expectation. Based on the two studies, different weather and sun radiations are shown at the thermal comfort of the container house. Therefore, designers could acquire more adequate information for evaluating and designing the HVAC in a preliminary design stage via the efficient simulation process.

Keywords: thermal comfort; admittance method; container house; simulation process

Applied System Innovation – Meen, Prior & Lam (Eds)
© 2016 Taylor & Francis Group, London, *ISBN 978-1-138-02893-7*

Greenness and livability: An interwoven approach to successful environmental planning

H.T. Tang & Y.M. Lee
Institute of Natural Resources Management, National Taipei University, New Taipei, Taiwan

ABSTRACT: To realize sustainable environmental planning, an interwoven approach that addresses both Greenness and Livability concerns should be adopted. The application of informed decision making process to sustainable urban development is the objective of this study, which uses a combination of decision making models and indicator development techniques to form a logical sequence of steps and a clear set of directions to assist decision and policy makers to identify solutions that benefit all stakeholders.

Keywords: environmental planning; sustainable urban development; informed decision making; greenness; livability

Innovation Design & Creative Design

Applied System Innovation – Meen, Prior & Lam (Eds)
© 2016 Taylor & Francis Group, London, *ISBN 978-1-138-02893-7*

Angle deviation method for measuring the number of bubbles

Ming-Hung Chiu & Jian-Ming Huang
Department of Electro-Optical Engineering, National Formosa University, Huwei, Yunlin, Taiwan

ABSTRACT: We use the angle deviation method with the techniques of common-path heterodyne interferometry and surface plasmon resonance to measure the number of bubbles.

Applied System Innovation – Meen, Prior & Lam (Eds)
© 2016 Taylor & Francis Group, London, ISBN 978-1-138-02893-7

An innovation aid for developing products leading to LOHAS

Li-Hsing Shih
Department of Resources Engineering, National Cheng Kung University, Taiwan, R.O.C.

ABSTRACT: This study proposes an aid for facilitating product design, leading to LOHAS (Lifestyle Of Health and Sustainability) by integrating persuasive technology and case based reasoning. A case library includes 98 existing products, in which users have been persuaded to a more human experience for fun, health, or energy harvesting. For each case, information that relates to a domain model covering target behaviors, users' motives, users' enhanced ability, design principles, and applied technologies are extracted and stored. By following the concept of case based reasoning, a conditional search has been conducted to retrieve cases and get useful suggestions of design principles and applicable technologies by inputting target behaviors and target customers' motives and abilities. A computer program is written to expedite the conditional search and retrieve useful cases from the case library. The designer can either be directly inspired by referring to retrieved cases or start with suggested design principles and applicable technologies for achieving an innovative product design.

Applied System Innovation – Meen, Prior & Lam (Eds)
© 2016 Taylor & Francis Group, London, *ISBN 978-1-138-02893-7*

Relationships between aesthetic preferences, affective quality, and feeling quality of graphic user interfaces

Shih-Miao Huang
Department of Mechanical Design Engineering, National Formosa University, Hu-Wei, Yunlin, Taiwan

Shu-Chu Tung
Department of Environmental Engineering, Kun Shan University, Tainan, Taiwan

Wu-Jeng Li
Department of Mechanical Design Engineering, National Formosa University, Hu-Wei, Yunlin, Taiwan

ABSTRACT: This study argues that both the affective qualities and feeling qualities of the system interfaces would be factors influencing a user's aesthetic preferences. Affective quality is a set of affective terms related to emotional responses. Feeling quality is another set of all the feeling terms to describe the feelings represented from skins. A path analysis was performed to explore the relationship between aesthetic preference, affective quality, and feeling quality. Sixteen Windows media player skins from the Windows official website were rated with aesthetic preferences, affective quality, and feeling quality. The outcomes indicate that affective quality did not significantly affect aesthetic preferences, but affected feeling quality. There were seven feeling quality terms selected in the aesthetic predicted model. They were *Delicate*, *Hi-tech*, *Formal*, *Fierce*, *Unique*, *Tight*, and *Robust*. The outcomes imply that the skin appearances with the feelings of "Hi-tech", "Formal", "Fierce", "Unique," and "Robust" were well received. A delicate appearance of the interactive skins was most important to design an aesthetic skin, but the skins with a tight feeling were considered to be the worst. It suggested that the designers have to design a skin with a delicate feeling, but avoid a skin design with a tight feeling.

Applied System Innovation – Meen, Prior & Lam (Eds)
*© 2016 Taylor & Francis Group, London,**ISBN 978-1-138-02893-7*

Battlezone-Mapping game space into the museum

Pey-Yune Hu
Department of Motion Pictures and Video, Kun Shan University, Tainan, Taiwan

Pei-Fang Tsai
Department of Public Relations and Advertising, Kun Shan University, Tainan, Taiwan

ABSTRACT: This paper presents a series of interactive media designs for the exhibition project of the National Science and Technology Museum. The exhibition contents are transformed by using interaction through the medium of games. Unlike the traditional exhibition space design of the museum, this project has mapped the special features of virtual video game spaces onto physical exhibition design planning, and it has also added role-playing and game elements to change the manner of visiting the exhibition. This paper has principally discussed the impact of the virtual space concept on physical space design caused by the development of digital technology. It has simultaneously explored the development process of the major design concepts belonging to this project, as well as introduced interactive installation specially designed for this exhibition. The aim is to transmit—knowledgably—information through an edutainment approach and enhance the participation initiative of visitors, thus, allowing them to maintain their interests throughout the visit.

Applied System Innovation – Meen, Prior & Lam (Eds)
© 2016 Taylor & Francis Group, London, ISBN 978-1-138-02893-7

Exploring the impact of design innovation on the enhancement of brand strength—a case study of the creative industry in Taiwan

Kuen-Meau Chen & Ying-Sin Lin
Lienda, Miaoli, Taiwan, R.O.C.

ABSTRACT: The major purpose of this study is to explore the efforts of the OBMs of the creative industry in Taiwan at improving their design-driven innovation and to investigate the effects of brand awareness. The findings of this study are expected to help the said creative industry to obtain a good understanding of the value of original brands. First, five OBMs of different types from Taiwan's creative industry are studied through in-depth interviews conducted with experienced supervisors from each OBM. Their views and opinions on design-driven innovation and brand strength are explored. Second, a literature review is conducted, and verbal analysis is performed on the in-depth interviews and factors that affect design-driven innovation and brand strength, which are then summarized. Results of the study reveal new sources of design-driven innovation, which consist of the following three new sub-dimensions: (1) design department in the value chain inside the company, (2) independent designers or design companies in the value chain outside the company, and (3) raw material suppliers and other factors in the value chain outside the company.

Applied System Innovation – Meen, Prior & Lam (Eds)
© 2016 Taylor & Francis Group, London, *ISBN 978-1-138-02893-7*

Research on the application of calligraphy elements in product design

Fu-Yuan Li
Department of Industrial Design, Tatung University, Taipei, Taiwan

ABSTRACT: Various forms of Chinese arts and cultural expression include painting, sculpture, music, and all kinds of performing activities, all presenting Chinese life experience and values. Among them is Chinese calligraphy—a form of art that represents unique aesthetics with structured shapes to express artistic conception. Nevertheless, with the advancement in modern digital technology and inconvenience in doing calligraphy, the art has been gradually disregarded. Therefore, this paper is set out to revitalize the value of Chinese calligraphy by specifically analyzing the way calligraphy is created and applying its creative elements to product design.

Through research on the form and creativity of calligraphy, as well as interviews with renowned calligraphers in Taiwan, this paper not only develops the design specifications of products, but also verifies the feasibility of design specifications. The result shows that in applying elements of Chinese calligraphy, one should stress the metaphysical spirit embodied in writing rather than in its physical structure. Additionally, when making design applications, more raw materials are recommended to be used in order to highlight the characteristics of the brush and ink. Moreover, in the process of transforming calligraphy into product design, one needs to consider the cognitive ability of the public towards the history of writing in order not to turn designing into a loner's game.

Keywords: chinese calligraphy; design specifications

Applied System Innovation – Meen, Prior & Lam (Eds)
© 2016 Taylor & Francis Group, London, ISBN 978-1-138-02893-7

A study of the usability of smartphone's main menu for seniors

N.T. Wang & M.C. Lo
National Taiwan University of Arts, New Taipei City, Taiwan

H.C. Tang
Vanung University, Taoyuan, Taiwan

ABSTRACT: Smartphones are commonly used by seniors nowadays. However, the weakening of their senses makes it difficult to operate them. In order to find out the usability of smartphones on the basis of how the seniors are reacting to different smartphone interfaces, three types of interfaces were tested in 102 seniors, aged between 45 and 65 years old. The result shows that the interface with the highest usability was the Grid-type, followed by the Horizontal-type, and the Irregular-type had the lowest usability.

Applied System Innovation – Meen, Prior & Lam (Eds)
© 2016 Taylor & Francis Group, London, ISBN 978-1-138-02893-7

Smart navigation: A study on iBeacon technology used in a digital tour of the museum

Pai-Ling Chang
Department of Digital Multimedia Arts, Shih Hsin University, Taipei, Taiwan

Yu-Lin Hsu
Department of Visual Arts, National Pingtung University, Pingtung, Taiwan

ABSTRACT: This study is focused on the iBeacon technology and how to use it as an interface for museum digital navigation design. By comparing the traditional audio tour and multimedia tour onto the user experience in the museum exhibition, and offer a new smart phone tour model, this study suggests that iBeacon positioning system can enhance the interactive experience of the viewer and the works. The meanings of the art work in the museum exhibits can effectively convey to the viewer. The research topics include: (1) interactive experience for digital tour, (2) the differences between smart phone tour and the traditional audio tour, (3) the design concepts of the smart phone tour. The conclusions can provide to museum as reference in digital navigation and exhibitions planning.

Applied System Innovation – Meen, Prior & Lam (Eds)
© 2016 Taylor & Francis Group, London, ISBN 978-1-138-02893-7

Fabrication of micro lenses by steel ball indentations

Meng-Ju Lin & Yu-Ru Lai
Department of Mechanical and Computer Aided Engineering, Feng Chia University, Taichung, Taiwan

ABSTRACT: Micro lenses used in MEMS are often fabricated by silicon-based micromachining and need expensive processes. A novel method to fabricate micro lens by indentation is investigated in this work. In this work, it is shown that micro lenses can be successfully fabricated by indentation methods. This method has advantages of simple processes and equipment, fast and massive fabrication, and does not require complex design and analysis. Therefore, it has practical potential. Four diameters of steel balls are used. The indentations of steel balls on aluminum plates can be used as molds to fabricate micro lenses. The micro lenses are composed of plastic materials. They have efficient focus ability. The experimental results indicate that the curvature radius of mold of indention on aluminum plates is coincident with design. It is found that micro lenses made from indentation of larger steel balls would have a better round shape and smaller roughness.

Applied System Innovation – Meen, Prior & Lam (Eds)
© 2016 Taylor & Francis Group, London, ISBN 978-1-138-02893-7

Value-added applications for the integration of dot-matrix hologram's iridescent effects and movable type printing technology

Yi-Ting Tsai, Hsi-Chun Wang & Yu-Lan Chiu
Department of Graphic Arts and Communications, National Taiwan Normal University, Taipei, Taiwan

ABSTRACT: Since the traditional movable type printing technology was replaced by computerized typesetting, it has turned out to be a historical artifact. The purpose of this research is to convert this movable type printing technology into a personal value-added signet with anti-counterfeiting capabilities using a dot-matrix hologram. The iridescent effect of the hologram is also quantitatively analyzed. The method proposed will have some potential in developing a unique value-added application for product protection.

Applied System Innovation – Meen, Prior & Lam (Eds)
© 2016 Taylor & Francis Group, London, ISBN 978-1-138-02893-7

A novel RFID-based wireless thermal convection angular accelerometer for environmental safety monitoring

Jium-Ming Lin
Department of Communication Engineering, Chung-Hua University, Hsin-Chu, Taiwan, R.O.C.

Cheng-Hung Lin
New Ph.D. Program in Engineering Science, College of Engineering, Chung-Hua University, Hsin-Chu, Taiwan, R.O.C.

Chia-Hsien Lin
Department of Electrical Engineering, Chung-Hua University, Hsin-Chu, Taiwan, R.O.C.

ABSTRACT: This paper applied five ideas to integrate an RFID tag with a thermal convection angular accelerometer on a flexible substrate. The first idea is that it is without any movable parts and very reliable. The second idea is to make it on a flexible substrate to save power, e.g., plastic or polyimide, since the thermal conductivity of those is much lower than silicon. The third idea is to apply xenon gas in a chamber instead of the previously used carbon dioxide to avoid oxidation effect on the heater and thermal sensors. The fourth idea is to integrate the angular accelerometer with an RFID tag on the same substrate to become a wireless sensor, and it is very easy to deploy it in the wild fields to monitor bridges, mudflows, and landslides. The final new idea is to make the thermal sensors in a non-parallel structure instead of the previous parallel one. The sensitivity of the latter is larger (0.0896 K/(rad/s^2)) but with a nonlinear effect at a smaller angular acceleration. However, the sensitivity of the former is smaller (0.0833 K/(rad/s^2)) but more linear.

Applied System Innovation – Meen, Prior & Lam (Eds)
© 2016 Taylor & Francis Group, London, ISBN 978-1-138-02893-7

The development of the asymmetric hexa-rotor aerial vehicle

Jie-Tong Zou, Chien-Yueh Hsu & Rui-Feng Zheng
Department of Aeronautical Engineering, National Formosa University, Huwei, Yunlin, Taiwan

ABSTRACT: The technology of Unmanned Aerial Vehicles (UAV) has been developing very fast during these last years. The multi-rotor Vertical Take-Off and Landing (VTOL) UAV, which can fly stable and hover in a fixed position, was developed rapidly. The multi-rotor UAV has many advantages: it has a simple mechanism, is safer than a helicopter, it has VTOL ability, is of small size and agile maneuverability. These multi-copters have many applications and can fly indoors and outdoors.

In this research, we developed an asymmetric hexacopter (Flyduspider), which has the following advantages: it has an easily identifiable heading angle and larger payload ability. The dynamic model of the hexacopter has been investigated in this research. During the developing process, we performed many simulations and experiments for choosing a suitable motor and propeller. These results can help us to design and build energy-saving multi-rotor aerial vehicles with a high efficiency and long endurance time.

After building the hexacopter, we started to proceed with a series of functional tests such as altitude hold flight, hovering (Loiter), return to launch, and autopilot.

Applied System Innovation – Meen, Prior & Lam (Eds)
© 2016 Taylor & Francis Group, London, ISBN 978-1-138-02893-7

Printmaking in a new area

Qi Luo
Drawing and Painting, College of Art, Guizhou University, China

Kuo-Kuang Fan & Ming-Chyuan Ho
Graduate School of Design, National Yunlin University of Science and Technology, Taiwan

ABSTRACT: The printing technique has become more diverse since the 20th century. Printmaking artists promote the development of new techniques. Such new techniques consist of two or more techniques rather than one. Before the reproduction of photography in the 1960s, the printing technique was more and more complicated. However, it now inevitably approaches the Chinese contemporary art. We present this report in order to make such a deep-rooted culture of art more widespread. Considering art printing from all points of view., we ought to learn from experience to explore every possibility: how to meet the requirements of the times and how to open a start-up.

Applied System Innovation – Meen, Prior & Lam (Eds)
© 2016 Taylor & Francis Group, London, ISBN 978-1-138-02893-7

Wind driven mechanism for solar-panel cleaning

Song-hao Wang, Syun-Cheng Lin & Yu-Ching Yang
Department of Mechanical Engineering, Kun Shan University, Tainan, Taiwan

ABSTRACT: For the benefit of the earth and the human beings, development and application of green renewable energy has become more and more important. At present, the first choice of green renewable energy is solar power generation systems. Solar power generation units are installed in the livelihood and public facilities, and have become very popular. Recent researches found that the level of power generation efficiency of solar cells is significantly affected by the dust on the surface. For example, four grams of dust on one square meter solar panel could reduce 40% of the solar-power generated. Therefore, how to keep the solar cell surfaces clean is a very important task.In order to clean the surface of the solar panel, a wind driven mechanism has been developed. The working principle is: a bi-directional reciprocating linear cam is applied to transform the rotation into a linear movement. Therefore, the wind drives the turbine and the turbine drives the cleaning brush. Compared with the devices currently available in the market, which use electrical motors, this mechanism is electricity free. In this paper, a low starting wind speed turbine is also discussed.In addition to cleaning the solar panel's surface, the wind turbine could be used mainly for generating the electric power. Whenever the solar panel needs to be cleaned, a simple clutch is enough to connect the wind turbine to complete the surface cleaning task.

Keywords: solar panel; cleaning; wind turbine; reciprocate cam.

Applied System Innovation – Meen, Prior & Lam (Eds)
*© 2016 Taylor & Francis Group, London,**ISBN 978-1-138-02893-7*

The imperative of congruent vision: Aligning the vision on innovations in SMEs

Kim C.K. Lee
Department of Industrial Design, Chaoyang University of Technology, Wufong, Taichung, Taiwan

ABSTRACT: It has been a significant trend for almost every discipline to rush into innovations recently. In order to help industrial designers to keep their reputation as professional innovators, this research began an exploration of how industrial designers are doing in innovations. From 2013 to 2015, 40 in-depth interviews with industrial designers in Taiwan were conducted. The result showed that most of them were deeply frustrated since their managements were too conservative in innovations. Actually, it will be a good chance for them to initiate design-led innovations if they could bring their companies to thrive through innovations. To help them to achieve this goal, this research proposed a series of papers that aimed at offering a model for these design-ers to formulate competitive design strategies. This paper is the second in that series, which was aimed at introducing the importance of "Vision Alignment" before design strategies could be formulated effectively.

Applied System Innovation – Meen, Prior & Lam (Eds)
© 2016 Taylor & Francis Group, London, *ISBN 978-1-138-02893-7*

Difference analysis of users' needs in automobile interior for female office workers–compact cars as example

Jui-Che Tu
Graduate School of Design, National Yunlin University of Science and Technology, Taiwan

Ya-wen Tu
Department of Commercial Design, Chienkuo Technology University, Taiwan
Graduate School of Design, National Yunlin University of Science and Technology, Taiwan

Yu-Ting Hung
Graduate School of Creative Design, National Yunlin University of Science and Technology, Taiwan

ABSTRACT: To understand female office workers' true needs regarding an automobile's interior decoration, this study adopted the case interview method in the first phase to explore the purchase trend of female buyers of automobiles of the best-selling brands in Taiwan and their views on automobile interior decoration. Then this study summarized this information into an industry-end interview summary as the basis for a questionnaire design. In the second phase, a questionnaire survey was conducted to explore the female office workers' lifestyle features and their automobile interior decoration requirements. Using SPSS, this study analyzed the differences in the automobile interior decoration requirements among female office workers of different occupations. Finally, the design principles for automobile interior decoration that were most suitable for this target group were concluded.

Applied System Innovation – Meen, Prior & Lam (Eds)
© 2016 Taylor & Francis Group, London, *ISBN 978-1-138-02893-7*

Strategies for revitalization of community craft industry

Po-Lun Hou & Ming-Chyuan Ho
National Yunlin University of Science and Technology, Yunlin, Taiwan

ABSTRACT: In recent years, the policy of the Ministry of Culture in community empowerment is gradually aiming at regional revitalization or industrialization. Many communities have begun to pay attention to their own culture or community characteristics and expecting to be "revitalized" or "transformed" or other expressions expecting the innovation of the community industry. This study reveals that when the community association turns toward the business-oriented community to achieve certain economies of scale, it will turn into a community of professional management of industrial section. The development of the industry in a community has a certain context and procedures; each stage of development priorities is different, but still follows a pattern of progressive development. In this study, five findings resulted after the intervention of crafts into the community. Through policies to help, it is easier to upgrade the community industry and the community cultural values can obtain a better preservation and activation as well.

Keywords: community craft, policy design, community industry

Applied System Innovation – Meen, Prior & Lam (Eds)
© 2016 Taylor & Francis Group, London, ISBN 978-1-138-02893-7

Research on teaching strategies to integrate creativity case activities into design courses

Tsen-Yao Chang
Department of Creative Design, National Yunlin University of Science and Technology, Yunlin, Taiwan

ABSTRACT: Teaching methods and models need to be adjusted and changed according to different students' growth and employment environments. In the designing field, with the division of professional work, talents in different fields should have interdisciplinary talents, such as communication and management bridge, while performing their own functions. Teaching strategies and methods respond to the demand of integrated design talents and cultivate integrated creativity talents with industrial ideas. Therefore, this study mainly explores creativity methods and leads creativity case activities into integrated design courses to conclude practical teaching strategies. Based on a course led by practical designs, this study takes a creativity case activity-special showcase exhibition as the center of the course. In this course, students must make an overall operation from a plain draft, via a three-dimensional model to a space planning and must think about color, structure, shape, and other conditions to finish the design program with a team of 5–6 people. In the course, participation and non-participation observation methods and course feedback sheets were applied to understand their learning performance and ideas in different stages of the courses by coding analysis. Furthermore, the Plan-Do-Check-Action (PDCA) structure was added to the action research theory. The creativity case activity-led integrated design course was divided into six contents for discussion consisting of: plan, do, observe, check, reflect, and action. Six indexes reflect introspections and feedbacks of the teaching contents and tasks of students in different stages. They were used as a basis for teachers to amend and adjust their teaching directions and further think about how to lead creativity activities into course teaching so as to increase the students' learning strength. In this way, this can be used as a reference of integrated design course planning in the future to cultivate talents with integrated design ability.

Applied System Innovation – Meen, Prior & Lam (Eds)
*© 2016 Taylor & Francis Group, London,**ISBN 978-1-138-02893-7*

Learning through doing—building successful engineering design teams using personality typing strategies

S.T. Shen
Department of Multimedia Design, National Formosa University, Hu-Wei, Taiwan, R.O.C.

Stephen D. Prior
Aeronautics, Astronautics and Computational Engineering Design, Faculty of Engineering and the Environment, University of Southampton, Southampton, UK

ABSTRACT: Engineering students learn to work best together using real-world exercises and challenging design briefs. The problem of who works best with whom and what type of personality is best to lead such a team is not well utilized in academic institutions. This paper seeks to present some new theories and concepts, which are realizable within an engineering educational context, to improve team working and inter-personal skills, as well as provide successful outcomes to projects. The main focus is on engineering design through Design, Make and Test (DMT), rather than design thinking. A new module entitled 'Systems Design and Computing', developed at the University of Southampton, is being used to improve Engineering Design Education using these tools and techniques.

Industrial Design & Design Theory

Applied System Innovation – Meen, Prior & Lam (Eds)
© 2016 Taylor & Francis Group, London, ISBN 978-1-138-02893-7

Intuitive interface design for elderly-demented users

Ya-Wen Cheng & Li-Hao Chen
Department of Applied Arts, Fu Jen Catholic University, New Taipei City, Taiwan

Yi-Chien Liu
Neurology Department, Cardinal Tien Hospital, New Taipei City, Taiwan

ABSTRACT: Dementia will prevent patients from easily using products in their daily life. Strengthening the intuitive interactivity of the product design interface will help elderly demented patients to live independently. This study aimed to explore the needs of elderly-demented users on the intuitive interaction operation affordance features of life product operating interfaces. Taking elderly-demented users as the main research target, this study collected different types of television remote controls as research vehicles and analyzed the direct response, complete mission time, video observations, and other data of subjects on simulation operating interfaces in order to understand the intuitive interactivity of product interfaces. The study results showed that the up and down volume buttons on a television remote control should be set independently; the model should be simple, free of text at the pressing area, and presented in the form of indicating symbols, so that elderly-demented users would be more able to perceive the intuitive operation affordance.

Applied System Innovation – Meen, Prior & Lam (Eds)
© 2016 Taylor & Francis Group, London, ISBN 978-1-138-02893-7

A TRIZ-based approach to integrating evolutionary trends and contradiction analysis for product design

Tien-Lun Liu & Fong-Kai Liang
Department of Industrial Engineering and Management, St. John's University, New Taipei City, Taiwan

ABSTRACT: Products are improved and designed to increase their ideality as defined in TRIZ. The design requirements are usually multidisciplinary and may lead to potential conflicts which have to be further resolved. In this research, we construct an analytical model by combining the concepts of evolutionary trends and contradiction analysis in TRIZ theory to explore the product design. Based on the evolutionary benefits from the progressing stages of each trend, we may relate the differences into 39 engineering parameters defined in TRIZ to acquire critical factors among components. The relationships are weighted with three levels: *Absolutely Relevant*, *Probably Relevant*, and *Probably Irrelevant*. The design contradictions could be further identified to resolve later. A monitor design example is provided to demonstrate the process of analysis. Such methodology is systematic to analyze the product contents by realizing the essence of the product design.

Applied System Innovation – Meen, Prior & Lam (Eds)
© 2016 Taylor & Francis Group, London, ISBN 978-1-138-02893-7

Children's intuitive and post use assessment of electronic drawing pens made for children

Tung-Hsun Chen
Department of Industrial Design, Tatung University, Taipei, Taiwan

Pei-Jung Cheng
Department of Media Design, Tatung University, Taipei, Taiwan

ABSTRACT: The prevalence of information technology has led to the development of electronic touch products, which has had a marked impact on children's education, learning, and lives. One such product is the electronic drawing pen: a stylus that enables children to draw directly on mobile devices. Regarding the shape and style of the drawing pen, those typical of ordinary pen were adopted. However, appropriate design of the drawing pens for children must consider their usage and habits; therefore, an intuition test and human factors and ergonomics test were conducted to facilitate designing the optimal pen shape and style for children to hold. In designing the tests, the shape and style of currently available drawing pens were analyzed, and three test samples of varying shape (i.e., triangular, quadrangular, and circular) were produced using a 3D printer. Interviews and observational analysis were undertaken to test 117 elementary school students aged 9–11 years old.

The study results are as follows: (1) the intuition test revealed that a drawing pen with a round barrel was the most suitable shape for the children, whereas a quadrangle barrel was the least suitable shape. Such a result might be attributable to drawing pens with a round barrel being commonly used by children and were smooth; (2) the pen-holding experiments also showed that drawing pens with a round barrel were the most preferred shape by all children in this study, indicating that the children intuitively selected drawing pens of a shape and style identical to the pens they had previously used; and (3) the image-tracing test showed that the children using the drawing pens with a round barrel needed to adjust their pen-holding position fewer times than when using the quadrangular or triangular drawing pens.

Applied System Innovation – Meen, Prior & Lam (Eds)
© 2016 Taylor & Francis Group, London, ISBN 978-1-138-02893-7

Design and experiments on the operating force of a sliding door

Wan-Fu Huang
Graduate Institute of Design Science, Tatung University, Taipei, Taiwan

Shih-Bin Wang & Chen-Hsi Huang
Graduate Institute of Design Science, Tatung University, Taipei, Taiwan
Department of Innovative Product Design, Lee-Ming Institute of Technology, New Taipei City, Taiwan

Chih-Fu Wu
Graduate Institute of Design Science, Tatung University, Taipei, Taiwan

ABSTRACT: The lack of operating force standards and measurement methods makes a universal design of doors difficult. In this study, the force measuring the system for a sliding door has been developed, and thus, the force measurements with and without a door closer were especially conducted. The results demonstrated that the Required Operating Force (ROF) for the sliding door tends to increase significantly as the door closer is applied. The force signals are characterized by two signals, composed of the Initial Operating Force (IOF) and ROF, and their difference becomes larger, especially at a slower door operating speed. The results showed that the use of the door closer not only makes the door's operation more complex, but also increases the burdens, particularly for people with poor individual capabilities. Furthermore, the results of the operating force signals revealed that the developed measuring methods can provide an effective evidence of the certification and examination of the ROF so as to complete the universal design of the sliding door.

Keywords: force measuring system; sliding door; required operating force; door closer

Applied System Innovation – Meen, Prior & Lam (Eds)
© 2016 Taylor & Francis Group, London, *ISBN 978-1-138-02893-7*

Customized design of female high-heeled shoes based on market study of consumer's demands and preferences

Hung-Jen Liao, Chien-Cheng Chang & Wen-Shung Ying
Department of Industrial Design, National United University, Taiwan

ABSTRACT: Customized individual commodities are sprouting in response to the growing demand of market. There are many types of customized merchandises. The design and development of women's products must meet the needs and preferences of the female consumer group. The main purpose of this study is to explore the design specification of female high-heeled shoes from the user group's purchase intention. Therefore, in order to understand the customers' voices, there should be an effective design communication, allowing customers to fully express their demands. This is the key to the overall success of the products before customized output is actually designed. Initially, a questionnaire survey was conducted for data collection. The female users at the Farmers Association at Miaoli, Gongkuan, were recruited for the survey of customized designing of female high-heeled shoes. The questionnaire survey contained female user's preferences towards the style of high-heeled shoes, frequency and occasions of using the product, problems in using high-heeled shoes and the source of information they use to purchase the product. Statistical analyses were then conducted to examine the special needs and differences among female users in the area. The data obtained from the quantitative analysis served as the set of guidelines for new design specifications for future design and development for female high-heeled shoes. At last, the female users at Miaoli, Gongkuan area were used as the target user group for the new high-heeled design and development.

Applied System Innovation – Meen, Prior & Lam (Eds)
© 2016 Taylor & Francis Group, London, ISBN 978-1-138-02893-7

A study of suitable adjective pairs for measuring tactile sensations

Wei-Ken Hung, Meng Hsuan Lin & Chien-Cheng Chang
Department of Industrial Design, National United University, Taiwan

ABSTRACT: This study examined the effectiveness of 10 adjective pairs for measuring the product's tactile sensations in tactile only as well as in the combined visual and tactile conditions. Two experiments (handling without seeing in a dark box, and handling with seeing) were constructed, and 30 participants were recruited to evaluate 6 different types of phone cases (diversified in terms of materials, textures, weight, and colors) on 10 adjective pairs (7-points scale) relative to the tactile sensations. The results exhibited that the evaluation value for each of the 10 adjective pairs to products in the "handling without seeing" experiment is not necessarily linearly correlated with the same adjective pair in the "handling and seeing" experiment, which means that the measuring for tactile sensations tends to be influenced by visual properties such as form, texture, and color. Three comparably stable adjective pairs between the two experiments were found; they are "luxury (economic-luxurious)", "warmness (cool-warm)", and "heaviness (light-heavy)" in order. Two significantly unstable and uncorrelated adjective pairs between the two experiments are "complexity (simple-complex)" and "roughness (smooth-rough)", while the other adjective pairs such as "uniqueness (typical-unique)", "regularity (irregular-regular)", "curvature (straight-curved)", "lumpiness (flat-lumpy)", and "hardness (soft-hard)" are in between. As mentioned above, although, the adjective pairs "complexity (simple-complex)" and "uniqueness (typical-unique)" tend to be influenced by the visual characteristics, we found that they are closely and linearly correlated to the "luxury (economic-luxurious)", in the "handling and seeing" as well as the "handling without seeing" experiments. The results show that the scales of complexity and uniqueness could be good determinants in predicting the material's luxury in different conditions, even when the participants would only touch the product without seeing it. The results could help researchers and designers to further understand the suitable adjective pairs for measuring tactile sensations in the visible and invisible conditions.

Applied System Innovation – Meen, Prior & Lam (Eds)
© 2016 Taylor & Francis Group, London, ISBN 978-1-138-02893-7

An integrated design approach based on conjoint analysis and TOPSIS algorithm to form design of product image

Hung-Yuan Chen
Department of Visual Communication Design, Southern Taiwan University of Science and Technology, Tainan, Taiwan

Yu-Ming Chang
Department of Creative Product Design, Southern Taiwan University of Science and Technology, Tainan, Taiwan

ABSTRACT: The commercial success of a product is dependent not only on its ease-of-use and functionality, but also on consumers' psychological response induced by the product form since a satisfactory product form can evoke consumers' pleasurable feeling or affective response. In this study, the electric shaver product is chosen for illustration purposes since it has broadly similar functional structures, and hence the consumers' response is governed primarily by its form or appearance. A series of experimental evaluation is conducted to collect evaluation results. An integrated design approach based on conjoint analysis and TOPSIS algorithm is proposed to establish the relationship between the product features of electric shaver and the corresponding consumers' responses. The proposed approach can assist product designers obtain the optimal combination of product features and provide the effective information so as to develop a new design candidate in terms of its ability to meet the demands of multiple consumers' psychological responses in the conceptual design stage of product form.

Applied System Innovation – Meen, Prior & Lam (Eds)
© 2016 Taylor & Francis Group, London, ISBN 978-1-138-02893-7

A study on psychology-based evaluation model for innovation design

C.T. Wu
Department of Industrial Design, National Kaohsiung Normal University, Kaohsiung City, Taiwan

L.F. Yang
Department of Information Communication, Kao Yuan University, Kaohsiung City, Taiwan

ABSTRACT: This research tries to explore psychology-based evaluation models for innovation design. In this article, the extensive QFD, developed by Wu, has been adopted for the procedure of innovative design. The major procedure of QFD is to identify the customers' needs for a product and then convert it into appropriate technical characteristics. According to the priorities of product characteristics, prior engineering parameters will be identified to become the key requirements for a redesign. The extension method will aid Customers' Requirements (CRs) to transform into product design attributes more comprehensively. For achieving an attractive design, we introduce the Kano model and EGM to construct the evaluation model. The proposed psychology-based evaluation procedure is mainly used in two parts of the innovative design. First, the evaluation process used in the QFD stage can help to identify attractive customers' needs. Second, the results of in-depth interviews help to discover the trend of innovative design. The flowchart of the proposed innovative design procedure with a psychology-based evaluation has also been developed.

Applied System Innovation – Meen, Prior & Lam (Eds)
© 2016 Taylor & Francis Group, London, *ISBN 978-1-138-02893-7*

Application of expert Delphi method in constructing evaluation indicators of design departments' competitive advantage

Shu-Ping Chiu & Li-Wen Chuang
Department of Digital Media Arts and Design, Fuzhou University, Xiamen, Fujian, China

Jui-Che Tu
Graduate School of Design Doctoral Program, National Yunlin University of Science and Technology, Taiwan

ABSTRACT: In Taiwan, the college enrollment rate has increased substantially. Taiwan has the highest percentage of university students in the world, but after industry migration and financial tsunami, newcomers in the workplace are facing a more arduous job-hunting situation. Therefore, the unemployment rate of college graduates has also gone up. This study aims at developing a set of evaluation indicators applicable to competitive advantage of design departments in Taiwan. This study was conducted in two stages. In the first stage, related literature, theories, and empirical researches were summarized; content analysis is used to draw the outline of expert interview and first draft of indicators. At the second stage, revised expert Delphi method is adopted to implement two rounds of expert questionnaire survey. In this way, evaluation indicators of design students' competitive advantages are constructed and the importance of the weight and sequence of these indicators are determined, thus, to provide emphasis and direction for design departments to plan the curriculum and evaluate teaching achievements, offer references for the diagnosis of design students' pre-service ability to design and understand that education is practical and applicable. Definite evaluation indicators developed here can also be used as a standard for related education units to assess students' competences, as well as bases for the industry to choose talents and make appraisal in the future.

Applied System Innovation – Meen, Prior & Lam (Eds)
© 2016 Taylor & Francis Group, London, ISBN 978-1-138-02893-7

From Tainan City attractions to explore the impression of color and sound

Ya-Ling Huang
College of Creative Media, Kun Shan University, Tainan, Taiwan

Bao-Feng Liao & Fan Hsu
Graduate School of Visual Communication Design, Kun Shan University, Tainan, Taiwan

ABSTRACT: This research program mainly surveys the need for urban color and sound impression in the "Smart Tour Guide City–Urban Demonstration Program" sponsored by the Department of Technology. The research selects four most typical tourist attractions along the route of No. 2 Bus and Green Line Bus of Tainan City, namely, Anping, Confucian Temple, Chihkan Tower and Xinhua Old Street Building and explores the impression of residents and tourists in color and sound at the four tourist attractions through questionnaire and interview. The survey result indicates that residents and tourists have a very similar impression in the sound and color at the four tourist attractions. Viewed from the color impression, the relationship between the main color and auxiliary colors has been identified at all the four tourist attractions. Related results will be provided to the urban demonstration program of the Department of Technology for use in the interactive design, color and sound planning relating to the urban tour guide design.

Cultural & Creative Research

Applied System Innovation – Meen, Prior & Lam (Eds)
© 2016 Taylor & Francis Group, London, ISBN 978-1-138-02893-7

Study on CPS spatial concept course and assessments for aboriginal children: With "rotation" and "mapping" as examples

Jen-Yi Chao & Lo Yi Yao
Graduate School of Curriculum and Instructional Communications Technology, National Taipei University of Education, Taipei, Taiwan

Chuan-Hsi Liu
Department of Industrial Education, National Taiwan Normal University, Taipei, Taiwan

Jen-Yang Chen
Department of Electronic Engineering, Ming Chuan University, Taoyuan, Taiwan

ABSTRACT: In order to develop a question-set assessment for daily spatial concepts with the CPS (Collaborative Problem Solving) approach for a Taiwanese aboriginal Atayal elementary school, this study designed an interactive digital material for the "rotation" and "mapping" units, with the support of three content experts and a pilot test. The result is a formal test paper for "rotation" and "mapping" concepts, comprising 8 question sets and 22 questions. The subjects of the study were 16 fifth-grade students from an Atayal elementary school in Nan'ao, Yilan. Data were collected from demonstrative lessons and tests. The result indicates that the fifth-grade students show better performance in the post-test than in the pre-test, after receiving CPS interactive digital demonstration lessons for the "rotation" and "mapping" concepts compared with the one received before.

Applied System Innovation – Meen, Prior & Lam (Eds)
© 2016 Taylor & Francis Group, London, ISBN 978-1-138-02893-7

A study on the key competencies of indigenous senior high students in PBL E-book production courses

J.Y. Chao
Graduate School of Curriculum and Instructional Communications and Technology, National Taipei University of Education, Taipei, Taiwan, R.O.C.

R.Y. Xu & Z.W. Jiang
Nan Oau Senior High School, Yilan, Taiwan, R.O.C.

ABSTRACT: This is a case study in which the researchers collaborated with Nan Oau Indigenous Senior High School and local craftsmen to conduct instructional courses in e-book production. This tripartite collaboration was formulated to encourage indigenous students to familiarize and interest themselves with local culture and to furthermore enhance their abilities in teamwork, communication, planning and other thematic oriented key competencies, such as information literacy. The course was conducted for two semesters for a total of 17 classes, 51 class hours, and was attended by 17 indigenous senior high students. The e-book production courses for both semesters each included the experiencing of local cultural activities, e-book production, and the display and discussion of completed works. Project-based learning (PBL) was utilized as the primary teaching strategy. Questionnaires were handed out before and after the courses to assess changes in the students' key competencies.

Applied System Innovation – Meen, Prior & Lam (Eds)
© 2016 Taylor & Francis Group, London, ISBN 978-1-138-02893-7

Study on application of seal cutting character art in costume design

Hsiu-Hui Hsu
Department of Science Design, Tatung University, Taipei, Taiwan
Department of Applied Cosmetology Director, LEE-MING Institute of Technology, Taipei, Taiwan

Yu-Chun Huang
Department of Science Design, Tatung University, Taipei, Taiwan

Wen-Chin Tsen
Department of Fashion and Design, LEE-MING Institute of Technology, Taipei, Taiwan

ABSTRACT: This study is designed to research the application of cultural background in costume design, where oriental elements and western modeling design are integrated and costume design is presented using new aesthetics of fashion design.

The sentence "The West Lake is beautiful after a rain in the awakening spring" in the poem of Ouyang Xiu, the Chinese literateur in the Song Dynasty, is the principle used. This poem mainly narrates the scenery of West Lake after the spring rain. This study used its artistic conception as the inspiration for the design and designed the poem's content into totems using seal cutting character art. Then, the totem design was created on fabrics with computer embroidery technology such that the fabrics display stereoscopic, delicate, and pierced visual effects through a characteristic zig-zag type layout design, color image expression as well as selection and use of fabric materials. Finally, the created fabrics were applied in fashionable garments, bags, and shoes to present the overall modeling design with a Chinese culture theme.

This study applied seal cutting art characters having cultural implications and beauty at the same time and poems in costume design and fashion design and presented them in different design manners, expecting to endow the fields of costume design and fashion design with a new style and texture.

Keywords: seal cutting character; fabric design; costume design; fashion design

Applied System Innovation – Meen, Prior & Lam (Eds)
© 2016 Taylor & Francis Group, London, ISBN 978-1-138-02893-7

Leading-in and promotion of enterprise safety culture—integrated with multimedia design

Yu-Lin Hsu
Department of Visual Arts, National Pingtung University, Pingtung, Taiwan

Pai-Ling Chang
Department of Digital Multimedia Arts, Shih-hsin University, Taipei, Taiwan

ABSTRACT: This study mainly presents the integrating multimedia design and aggregating specific measures of the enterprise's promoting safety culture, including the items such as the history of business growth, the awards and certification over the years, participation in government-related projects, and advancing safety-related activities. It is operated in the form of multimedia through the records of site visiting of various companies and customized security diagnostic systems, together with the measures of leading-in and promoting by recording, concluding and analyzing the enterprise's safety culture. It records and shares by applying multimedia presentation as its platform and then effectively leads the enterprise to a specific practice of safety culture, expecting to enhance the leading-in and advancement of enterprise safety culture and further promote the levels of enterprise safety culture so as to construct the enterprise's extraordinary safety image.

Applied System Innovation – Meen, Prior & Lam (Eds)
© 2016 Taylor & Francis Group, London, ISBN 978-1-138-02893-7

A study on the Kansei imagery survey model based on cultural differences

Chen-I Huang & Shing-Sheng Guan
Graduate Institute of Design, National Yunlin University of Science and Technology, Douliou, Yunlin, Taiwan, R.O.C.

ABSTRACT: This research first conducted a questionnaire, then compiled and analyzed the data from that questionnaire in order to understand the relationship between the results of the application of Kansei engineering and cultural elements:

Based on the previous research of the Kansei imagery model, the researchers have gained two different Kansei imageries—the "Eastern" and the "Western" – that can be used as a foundation for marketing purposes. The results of the questionnaire analyses on the "Western" respondents are thus: the researchers have used three samples of chocolate packaging to investigate how "Western" respondents perceive the Kansei imageries. These three samples consisted of "modern/traditional," "rational/emotional," and "Individual and popular" elements and these collocated elements showed significant differences.

The results of the questionnaire analyses on the "Eastern" respondents were: based on the "Western products" as questionnaire samples, it was found that there was a significantly different perception of Kansei imagery between the "Eastern" and the "Western" respondents.

Applied System Innovation – Meen, Prior & Lam (Eds)
© 2016 Taylor & Francis Group, London, ISBN 978-1-138-02893-7

Investigating models for reuse of historic buildings—Japanese-style dormitories

Szu-Yun Lai & Shang-Chia Chiou
Graduate School of Design, National Yunlin University of Science and Technology, Yunlin, Taiwan

ABSTRACT: In recent years, the preservation of historical buildings has been reused in different forms. There are many historical buildings from the Japanese colonial period in Taiwan, and they are facing preservation or demolition. Without proper management and maintenance, even if we exclude the environment, health and safety, historical buildings could not escape from discard, retaining not only building, but also the history of the building. Giving these unused spaces a historic significance and a new value is the way to attain community satisfaction. The purpose of this research is to explore how the Japanese-style dormitories of historic building can be reused. We can find that it is related to cultural heritage category and property ownership. The truth of reusing the unused historical building is to make more people interact with the space, enable the city to coexist with the new rising and the old in harmony, and enable people to identify the building.

Applied System Innovation – Meen, Prior & Lam (Eds)
© 2016 Taylor & Francis Group, London, *ISBN 978-1-138-02893-7*

The planning and art creation for cultural game development—an example of Donggang King Boat Ceremony

Dawei Lin
National Pingtung University, Pingtung, Taiwan

ABSTRACT: The Wang Ye worship is one of the representative Taoist beliefs in Taiwan diverse religious culture. It was originally the god of plague for helping people to get rid of plagues. After localization, Wang Ye worship has become the god who meets the people spiritual sustenance with multi-function blessing. The King Boat Ceremony is lively and spectacular, and it has become a Taiwan's famed religious culture. This research aims to develop a cultural game by interactive storytelling. The history survey and interview, therefore, is conducted for game planning and game art creation. A game entitled as King Boat Ceremony is created based on this meaningful history and interesting culture. This cultural game includes three levels: the first level describes the story of Wen Hung how to be the god of plague. The second level illustrates the god of plague how to protect people. The final level narrates the King Boat Ceremony. Through the fun gameplay, the player can discover the unforgettable story in Donggang, Taiwan.

Applied System Innovation – Meen, Prior & Lam (Eds)
© 2016 Taylor & Francis Group, London, ISBN 978-1-138-02893-7

Applying statistical analysis to deconstruct the transformation process of political campaigns in the information age

Sieng-Hou Chen
Doctoral Program, Graduate School of Design, National Yunlin University of Science and Technology, YunTech, Taiwan

Li-Hsun Peng
Department of Creative Design, National Yunlin University of Science and Technology, YunTech, Taiwan

ABSTRACT: The prevalence of mobile devices has rendered cyber culture a major innovative force worldwide. Online communities have actively participated in social affairs and street politics. From the Arab Spring protest in 2010 and the Occupy Wall Street movement in 2011, to Taiwan's Sunflower Student Movement in 2014, people of the younger generation have shown their abilities to create numerous miracles through their mobile devices, thus transforming these devices into an effective tool for election purposes. The Internet began to influence the form and content of Election Advertisements (EAs). However, such a change not only affected the production of EAs but also influenced the election results. This shows that an invisible information revolution is advancing its way across the world.

This study adopted visual analysis and examined the development process of the Internet and EAs, investigating how online media, communities, and users have transformed the form and content of EAs in Taiwan and exploring with such transformation entails. The study results indicated that previous EAs had emphasized the image and visibility, and were typically disseminated through common media means. By contrast, new-generation EAs diverged from the control of election camps. Through methods developed by public voters and Internet users, various novel EAs have been promulgated to the public. For future campaigns, the collective power of online groups will be further strengthened to an extent that campaign candidates cannot manipulate public opinions. All candidates will be dissected within the waves of public opinion. Such a phenomenon facilitates dissolving the confrontations between the Kuomintang and Democratic Progressive Party, thereby eliminating the black–gold political corruption that had plagued the politics of Taiwan for many years.

Applied System Innovation – Meen, Prior & Lam (Eds)
© 2016 Taylor & Francis Group, London, ISBN 978-1-138-02893-7

Study on cultural information transmission model for design of cultural products

Chi-Hsiung Chen
Department of Product Design, Chungyu Institute of Technology, Keelung, Taiwan
Graduate School of Creative Design, National Yunlin University of Science and Technology, Yunlin, Taiwan

Shih-Ching Lin
Doctoral Program, Graduate School of Design, National Yunlin University of Science and Technology, Yunlin, Taiwan

ABSTRACT: As Taiwan has long been influenced by foreign cultures, it has diverse rich and delicate cultures, and has accumulated strong cultural energy. Meanwhile, Taiwan's government also actively promotes policies related to the cultural and creative industry and further pushes forward the developmental trend of the creative industry. New opportunities for numerous cultural and creative products (hereinafter referred to as cultural products) are thus expanded, and they have been highly concerned by the academic circle and industry as well as have obtained great acceptance from the consumer market in recent years. After analyzing related documents, the study found that there was a considerably large difference between consumers' cognition of cultural products and meanings that designers intended to convey. Therefore, the study aims to probe into how cultural products transmitted their cultural information to consumers after cultural elements were reinterpreted through different design skills. By adopting case studies and field research, the study analyzed and evaluated the cases of the cultural and creative industry in Taiwan, and established reference models and indicators to provide references for the companies of the industry when they design products of this kind in the future. The results of the study are: (1) The value of Taiwan's cultural products could be effectively elevated through the introduction of craftsmanship. (2) The memory links of consumers could be effectively achieved when products were designed through consumers' experience. (3) Cultural information could be effectively transmitted to consumers, and new cultural styles could be created when cultural products were designed with a three-layer model including strategy layer, meaning layer and technique layer.

Applied System Innovation – Meen, Prior & Lam (Eds)
© 2016 Taylor & Francis Group, London, ISBN 978-1-138-02893-7

The signs of Taiwanese traditional gold jewelry designs

I.C. Chen & S.C. Chiou
Graduate School of Design, National Yunlin University of Science and Technology, Yunlin, Taiwan

ABSTRACT: Since ancient times, the Chinese like to apply traditional auspicious totems or characters on gold jewelry. On wedding days, the Taiwanese like to wear gold jewelry with auspicious signs to raise spiritual happiness. Since there is a variety of traditional auspicious totem designs, this research uses the Delphi Method to collect Chinese auspicious totems that are popular for weddings. The findings show that ten types of totems and characters are often used in weddings to represent the auspicious consciousness, mainly including animals, plants, and characters. The traditional auspicious totems contain thought-provoking meanings to convey prosperity, peace, and other traits. After exploring the sign value of traditional wedding's gold jewelry, this study constructs the spiritual value for design totems on gold products.

Keywords: traditional sign; gold jewelry; Delphi method

Applied System Innovation – Meen, Prior & Lam (Eds)
© 2016 Taylor & Francis Group, London, ISBN 978-1-138-02893-7

Study on the rhombic stripes of the aboriginal traditional apparels in Taiwan

Wang Po Hsun
Faculty of Humanities and Social Sciences, City University of Macau, Macau

ABSTRACT: The decorated stripes used on the Taiwanese aboriginal apparels contain both concrete and abstract graphs which constitute abstract geometric forms and which form grouped stripes. The rhombic stripe is a common design among the aboriginals frequently used on the embroidery products. The sum total is 73 items of the embroidery products of Atayal people collected from the museums at home and abroad as well as the private collectors. The methods of the paper are description, illustration and comparison; they would be then adopted to interpret the rhombic expression on the embroidery products. This was made to obtain the relation between the embroidery structure and the form from the perspective of construction design. The primary decorative theme on the Atayal grouped stripes is the rhombic stripe. After the analysis of the 220 rhombic stripes, we can find that there are different forms of expression among the stripes. This can be generated from both internal structure and external structure. For the internal structure, there are rhombic shapes, flat crossed shapes, stripes, oblique stripes, segmentation, crossed shapes, rice-mortar shapes, rectangles, combinations, etc. For the external structure, there are flat crossed shapes, square crossed shapes, long and narrow shapes, etc. In addition, the researchers discover that there exists a close relation among the rhombic stripes, and it was already embedded with a systematic concept, in terms of its form.

Applied System Innovation – Meen, Prior & Lam (Eds)
© 2016 Taylor & Francis Group, London, ISBN 978-1-138-02893-7

The visual communication of the city image: Tainan City's case

Wang Po Hsun & Xing Ya Long
Faculty of Humanities and Social Sciences, City University of Macau, Macau

ABSTRACT: Tainan City, a city in Taiwan, inherits the cultural context of the paste with the experience of landscape construction. Now this city attempts to employ the royal poinciana as the visual image of the city. While the public sector uses the royal poinciana as the design theme, the private sector tends to follow this trend, building a small-scale city image. This study analyzes the royal poinciana badge design of Tainan City, collecting 64 samples in total. The researchers discover that the graphic design of the badge plays the major part in terms of morphology, mainly with the use of abstract expressive form. In order to cope with the unit category of the royal poinciana design, the word design of some school sectors, governmental sectors, and commercial sectors all employs the sign design as their habitual design strategies. The elements of the design themes consist of two models: the name of the organization and the organization's image. The images or the figures of the royal poinciana, abstract or concrete, are all done in a simplified manner of contour silhouette. As the morphological expression is concerned, the royal poinciana situated in the center serves as the component of design elements. From the perspective of constructing a city image, this design model can modulate a basic frame of the royal poinciana, and, after following some guidance of design principles, it can construct an ordered city image of the visual expression.

Applied System Innovation – Meen, Prior & Lam (Eds)
© 2016 Taylor & Francis Group, London, ISBN 978-1-138-02893-7

Micro-enterprise of cultural creative brands: Construction of entrepreneurial strategy and operation pattern

Tsen-Yao Chang
Department of Creative Design, National Yunlin Yuniversity of Science and Technology, Yunlin, Taiwan

ABSTRACT: Modern people have gradually changed the dimension of life's demands. Apart from the basic physiological demands, they value self-realization more. Under such changes, the micro-enterprises keep in line with the trend of self-style and dream realization to create the entrepreneurs' blueprint of life's dreams in a limited scale. When micro-enterprises develop in the cultural creative industry, the emotional resonance and promotion of local distinctness add to the locality's care and promotion of micro-enterprises of cultural creative brands, besides the operational ideality. Moreover, due to the small size, most micro-enterprises have fewer than five employees, who are more efficient and flexible for strategy promotion. However, apart from the passion of the dream, rational thinking and planning are also required. Therefore, this study interviewed a total of 11 micro entrepreneurs in the cultural creative industry in the northern, central, southern, and eastern parts of Taiwan. Through a qualitative study method, it encodes and analyzes various factors of the respondents during the entrepreneurship process. In this way, we expect to understand the characteristics of the micro-entrepreneurs in the cultural creative industry and the overall dimensions of the operation, which can be taken as a reference for the micro-entrepreneurs stepping into the cultural creative industry. Using grounded encoding, the study analyzes the interview data of 11 entrepreneurs to construct the internal and external factors of micro entrepreneurship. The internal factors emphasize the conditions and ideas of the entrepreneurs, including the cultural background, attitude, and spirit. The external factors include the environmental situation and operational development. Moreover, it proposes the shaping and operation of the microcultural creative brand style, which can assist brand operators in overall thinking from internal to external dimensions. Besides, they can perform introspection and refer to directions in each stage, so as to further establish brand identity and style distinctness and finally achieve a sustainable operation of the brand.

Applied Mathematics

Applied System Innovation – Meen, Prior & Lam (Eds)
© 2016 Taylor & Francis Group, London, ISBN 978-1-138-02893-7

Driver workload evaluation using physiological indices in dual-task driving conditions

Gao Zhenhai, Li Yang & Duan Lifei
State Key Laboratory of Automobile Simulation and Control, Jilin University, P.R. China

Zhao Hui
State Key Laboratory of Vehicle NVH and Safety Technology, Changan Automobile Holding Ltd., P.R. China

Zhao Kaishu
No1. Affiliated Clinical Hospital of Jinlin University, Changchun, China

ABSTRACT: The driver state information is indispensible for the development and evaluation of an advanced driver assistance system. To propose a driver mental workload evaluation method, physiological data, and subjective scores were recorded in a moving-base driving simulator under a dual-task condition composed of a driving task and an auditory n-back task. The heart rate variability, skin conductance level, and respiration rates were extracted and examined as a measure of a driver's workload. Subjective scores used the NASA-TLX and indicated that the driver's stress level increases under the dual-task condition, and a significant variation of physiological indices were found in the experimental trials. The results indicated a correlation between the physiological indices and the driver's mental workload. In order to verify the effectiveness of a simulator study, a combined measure was then created, using a multiple regression method based on the physiological indices. The evaluation method developed in this study can be used in the design of an advanced driver assistance system and a human–machine interface.

Applied System Innovation – Meen, Prior & Lam (Eds)
© 2016 Taylor & Francis Group, London, ISBN 978-1-138-02893-7

Applying Quality Function Deployment (QFD) to improve the competitiveness of a manufacturing enterprise

M. Barad
Tel Aviv University, Israel

ABSTRACT: To improve its competitiveness, an enterprise has first to prioritize its strategic improvement needs. To deploy the strategic improvement needs to the action level, we use Quality Function Deployment, a *structured* and *consistent* framework. QFD is a product planning quality technique which extracts the customer *needs* or desires expressed in his/her own words, translates them into technical quality characteristics of a product planning, design, process and production. However, Quality Function Deployment is not limited to product planning; it is a multi-purpose technique which can be used for translating high level *improvement needs* of various systems to lower levels. Our virtual example propagates the improvement needs of a manufacturing enterprise from the strategic level to the action level.

Keywords: strategic planning; quality function deployment; improvement needs; concerns

Applied System Innovation – Meen, Prior & Lam (Eds)
© 2016 Taylor & Francis Group, London, ISBN 978-1-138-02893-7

Application of the Refined Lindstedt-Pioncare Method based on He's energy balance method to a strongly nonlinear oscillator

G. Ge & Z.C. Yun
School of Mechanical Engineering, Tianjin Polytechnic University, Tianjin, China

ABSTRACT: In this paper, a Refined Lindstedt-Pioncare Method (RLPM) based on He's energy balance method was proposed. A new expression of approximate solution of the forced strongly nonlinear Duffing oscillator, together with the Hamilton function of the system, was introduced and be taken in to account. Thus, steady state frequency, amplitude and phase are calculated simultaneously. Precise enough results were obtained. Numerical simulations were carried out to verify the method. The results were consistent with the analysis.

Applied System Innovation – Meen, Prior & Lam (Eds)
© 2016 Taylor & Francis Group, London, ISBN 978-1-138-02893-7

Choquet integral regression model based on Liu's second order multivalent fuzzy measure

Hsiang-Chuan Liu
Department of Biomedical Informatics, Asia University, Taichung, Taiwan

Hsien-Chang Tsai
Department of Biology, National Changhua University of Education, Changhua, Taiwan

Yen-Kuei Yu
Graduate Institute of Educational Information and Measurement, National Taichung University of Education, Taichung, Taiwan

Yi-Ting Mai
Department of Sport Management, National Taiwan University of Sport, Taichung, Taiwan

ABSTRACT: The well-known Sugeno's Lambda-measure can only be used for real data fit to sub-additive, additive, or super-additive fuzzy measures, which cannot be mixed with any fuzzy measure. To overcome this disadvantage, Grabisch extended the fuzzy density function from the first order to the second order, in order to propose his 2-additive fuzzy measure. We know that the 2-additive fuzzy measure is only a univalent fuzzy measure. Hsiang-Chuan Liu has proposed an improved multivalent fuzzy measure based on a 2-additive fuzzy measure, called Liu's second order multivalent fuzzy measure. It is more sensitive and useful than a 2-additive fuzzy measure, since it is a generalization of the 2-additive fuzzy measure. However, the fuzzy density functions of all of the above mentioned fuzzy measures can only be used for unsupervised data. In this paper, we have proposed the corresponding ones for the supervised data. In order to compare the Choquet integral regression model with P-measure, λ-measure, Liu' multivalent fuzzy measure, 2-additive measure, and Liu' second order multivalent fuzzy measure based on Liu's supervised fuzzy density function, the traditional multiple regression model and the ridge regression model, a real data experiment by using a 5-fold cross validation Mean Square Error (MSE) is conducted. Results show that the Choquet integral regression model with Liu' second order multivalent fuzzy measure has the best performance.

Applied System Innovation – Meen, Prior & Lam (Eds)
© 2016 Taylor & Francis Group, London, ISBN 978-1-138-02893-7

Fuzzy optimization based on the mixed-integer Memetic Algorithm

Y.C. Lin & Y.C. Lin
Department of Electrical Engineering, WuFeng University, Chiayi, Taiwan

ABSTRACT: Many real-world mixed-integer optimization problems inherit a more or less imprecise nature. If we take into account the flexibility of constraints and the fuzziness of objectives, the original mixed-integer optimization problems can be formulated as fuzzy mixed-integer optimization problems. Evolutionary Algorithms (EAs) are population-based global search methods. They have been successfully applied to many complex optimization problems. Memetic Algorithms (MAs) are hybrid EAs that combine genetic operators with local search methods. With global exploration and local exploitation, MAs are capable of obtaining more high-quality solutions. On the other hand, Mixed-Integer Hybrid Differential Evolution (MIHDE), as an EA-based search algorithm, has been successfully applied to many mixed-integer optimization problems. In this paper, a mixed-integer memetic algorithm based on MIHDE is developed for solving the fuzzy mixed-integer optimization problems. The proposed algorithm is implemented and tested on a fuzzy mixed-integer optimization problem. Satisfactory results can be obtained in the computational experiment. This demonstrates that the proposed algorithm is promising to handle the fuzzy mixed-integer optimization problems.

Applied System Innovation – Meen, Prior & Lam (Eds)
© 2016 Taylor & Francis Group, London, ISBN 978-1-138-02893-7

Integrating steps for solving second- and third-order nonhomogeneous linear ordinary differential equations

Fann-Wei Yang & Chien-Min Cheng
Department of Electronic Engineering, Southern Taiwan University of Science and Technology, Tainan, Taiwan

Mei-Li Chen
Department of Electro-Optical Engineering, Southern Taiwan University of Science and Technology, Tainan, Taiwan

ABSTRACT: The general solution $y(x)$ for second- and third-order nonhomogeneous linear ordinary differential equations can be solved by an integrated operation, which starts setting $y(x) = f(x)y_1$, where y_1 is a solution of their corresponding homogeneous equations. Then $f(x)$ can be obtained by substituting the assumed $y(x)$ and its derivatives into original nonhomogeneous equations. We tentatively call this the integrated order-reduction method. And the results thus obtained are verified to be the same as those by the method of variation of parameters. However, there still are some distinctions to be noted, for instance that the general solution $y(x)$ is acquired directly by multiplying $f(x)$ by y_1, while the latter must undergo three steps to form the same formula. Additionally, the imposed conditions employed by the latter are disused in our calculation. This method of order-reduction can further be applied to solve fourth- or even higher-order nonhomogeneous linear ordinary differential equations.

Keywords: nonhomogeneous linear ODEs; order-reduction; general solution; disuse imposed conditions

Applied System Innovation – Meen, Prior & Lam (Eds)
© 2016 Taylor & Francis Group, London, ISBN 978-1-138-02893-7

Fracture mechanics analysis of corroded pipeline with different corrosion conditions using computational simulation

H.S. Huang
Graduate Institute of Engineering Science and Technology, National Kaohsiung First University of Science and Technology, Kaohsiung, Taiwan, R.O.C.

S.Y. Hsia
Department of Mechanical and Automation Engineering, Kao-Yuan University, Kaohsiung, Taiwan, R.O.C.

Y.T. Chou
Department of Applied Geoinformatics, Chia Nan University of Pharmacy and Science, Tainan, Taiwan, R.O.C.

Y.C. Yu
Graduate Institute of Electrical Engineering, National Kaohsiung First University of Science and Technology, Kaohsiung, Taiwan, R.O.C.

ABSTRACT: This study aims to apply Finite Element Method and Boundary Element Method to analyze the effects of crack-like flaw for local corrosion of petrochemical pipes on predicting the fatigue life and crack extension. Real corroded pipeline is examined to measure the thickness and analyze the stress distribution. Non-destructive testing results and software analyses show that cracked oil pipes with local corrosion bear larger stress, mainly internal pressure, on the longitudinal direction than the circumferential direction. As a result, the maximal fatigue loading cycle of a circumferential crack is approximately 25 times higher than it of a longitudinal one. From the growing length and depth of a crack, the final aspect ratio of crack growth appears in 2.42–3.37 and 2.71–3.42 on the circumferential and longitudinal direction, respectively. The complete crack growth and the corresponding fatigue loading cycle could be acquired to determine the service life of the oil pipe being operated as well as the successive maintenance period and method so as to largely enhance the security and activation in a petrochemical plant.

Keywords: local corrosion; crack-like flaw; boundary element method; finite element method; fracture mechanics analysis

Applied System Innovation – Meen, Prior & Lam (Eds)
© 2016 Taylor & Francis Group, London, ISBN 978-1-138-02893-7

Thermal characteristics of region surrounding laser welding keyhole

C.Y. Ho & Y.H. Tsai
Department of Mechanical Engineering, Hwa Hsia University of Technology, New Taipei, Taiwan

Y.C. Lee
Department of Architecture, National Taitung Junior College, Taitung, Taiwan

ABSTRACT: This paper investigates the thermal characteristics of region surrounding laser welding keyhole. The thermal characteristics of region surrounding laser welding keyhole play an important role on the mechanical properties after welding. The welding keyhole produced by laser welding is assumed to be a paraboloid of revolution. The laser beam with Gaussian distribution of intensity irradiates on the keyhole. Considering the phase change at solid-liquid interface, this study utilized the finite difference method to numerically calculate the thermal fields in the solid and liquid region surrounding laser welding keyhole.

Keywords: thermal characteristics; laser welding; keyhole

Applied System Innovation – Meen, Prior & Lam (Eds)
© 2016 Taylor & Francis Group, London, ISBN 978-1-138-02893-7

Reversible data hiding using content-based scanning and histogram shifting

M.N. Wu
Department of Information Management, National Taichung University of Science and Technology, Taichung, Taiwan, R.O.C.

G.S. Chen
Department of Computer Science and Information Engineering, National Taichung University of Science and Technology, Taichung, Taiwan, R.O.C.

ABSTRACT: Recently, the reversible image data hiding method has been widely used. The reversible image data hiding method based on image histogram shifting and inverse S-scan prediction achieves a large hiding capacity with a low embedding distortion without considering the image edge and the texture characteristic. This paper proposes an image scanning method with a minimum cost spanning tree of reference image to find a content-based path based on the image edge and the texture characteristic. It can improve the hiding capacity and reduce the prediction error. By comparing the experimental results obtained from other histogram shifting-based methods, our method shows more hiding capacity and higher image quality, and achieves a higher embedding capacity at the same PSNR level.

Keywords: reversible data hiding; content-based scanning; histogram shifting

Applied System Innovation – Meen, Prior & Lam (Eds)
© 2016 Taylor & Francis Group, London, ISBN 978-1-138-02893-7

A study on efficiency improvement of ion implanter by FMEA and FTA approaches

C.W. Huang & T.S. Chen
Institute of Engineering Management, National Cheng Kung University, Tainan, Taiwan

ABSTRACT: The popularity of mobile communication device leads the age from desktops usage to portables. Many IC design houses develop new products with multi-functional and complex chips used in phones and tablets to fulfill the global consumer demand. This phenomenon makes IC foundries increase the capacity to satisfy the market needs by IC design houses. Therefore, a better utilization of the bottleneck machines for improving the capacity extension of foundry is the top priority task.

In this study, the FMEA approach with microperspective and the FTA approach with macro perspective are combined for the application of bottleneck machines. The combination of these two modes is called the Dual Failure Analysis (DFA) model. The FMEA method is used to process the risk evaluations of detection, severity and occurrence. On the other hand, the FTA method is applied to evaluate the abnormal problems of machines in the manufacturing process. Thus, the root causes can be found and improved.

After the implementing of the proposed DFA model, the troubleshooting steps easily build and maintain the ion implanter machines. As the characteristics of both models are complementary without conflict, we finally gained higher machine utilization and obtained better productivity in the manufacturing process.

Keywords: FMEA; FTA; DFA; ion implanter

Management Science

Applied System Innovation – Meen, Prior & Lam (Eds)
© 2016 Taylor & Francis Group, London, ISBN 978-1-138-02893-7

Research on the improvement of products' external appearance by using Failure Mode and Effect Analysis

Tian-Syung Lan, Pin-Chang Chen & Chia-Ming Chang
Department of Information Management, Yu Da University of Science and Technology, Maioli, Taiwan, R.O.C.

ABSTRACT: The Failure Mode and Effect Analysis (FMEA) method is a kind of reliability analysis tool of prevention rather than an afterward remedy, which is used in evaluating the new design or checking the progress of design, so as to reduce the failure rate of the product in production or its procedure. It is a systematic method that is used to identify and investigate the potential weakness of products or production procedures. This study adopts the FMEA method to discuss the Failure Effect phenomenon that the finished product has poor appearance, which is caused by processing factors such as cast pouring, modeling and laser, and ink sealing. According to the results of this study, the poor influences on the finished product's appearance in the production process is revealed. And more importantly, the priority of risk factors and weight values of potential failure caused is ranked, which can be taken as reference by the engineering personnel to reduce the quality appearance's abnormity and give suggestions on the items that urgently need improvement. As a result, effective and preventive measures can also be taken to reduce the abnormality cost in production as well as the faults step by step, so as to improve the quality and the product yield.

Applied System Innovation – Meen, Prior & Lam (Eds)
© 2016 Taylor & Francis Group, London, ISBN 978-1-138-02893-7

Using DEMATEL-based ANP in analyzing critical factors and the customer preferences of 4G mobile services

Yu-Lin Liao
Department of Business Administration, Chung Yuan Christian University, Taoyuan, Taiwan
Department of Business Administration, Chien Hsin University of Science and Technology, Taoyuan, Taiwan

Yi-Chung Hu
Department of Business Administration, Chung Yuan Christian University, Taoyuan, Taiwan

ABSTRACT: With the innovation of mobile communication and multimedia technology, 4G LTE can provide faster data rates than the current 3G networks. This paper applies a Multiple Criteria Decision Making (MCDM) model which combines the methods of the Decision Making Trial and Evaluation Laboratory (DEMATEL) and the Analytic Network Process (ANP) to evaluate the critical factors of consumers using fourth generation mobile communication service and explore the causal relationship between customer preferences in Taiwan. DEMATEL technique is used to build a Network-Relationship Map (NRM) among criteria and the ANP based on the NRM, the interdependence and feedback among criteria can be used as decision priority on development alternatives. The results enable the managers to adopt the appropriate actions in achieving competitive advantages.

Keywords: Analytic network process; decision making trial and evaluation laboratory; 4th-Generation; multiple criteria decision making

Applied System Innovation – Meen, Prior & Lam (Eds)
© 2016 Taylor & Francis Group, London, *ISBN 978-1-138-02893-7*

Suppliers selection model in the Liquid-Crystal-Display industry

Chia-Nan Wang & Han-Sung Lin
Department of Industrial Engineering and Management, National Kaohsiung University of Applied Sciences, Kaohsiung, Taiwan

Yen-Hui Wang
Department of Information Management, Chihlee Institute of Technology, Taiwan

Ya-Ru Lee & Ming-Hsien Hsueh
Department of Industrial Engineering and Management, National Kaohsiung University of Applied Sciences, Kaohsiung, Taiwan

ABSTRACT: In this paper, the Thin-Film-Transistor Liquid-Crystal-Display (TFT-LCD) is manufactured, which is the upstream of the backlight module industry. The industry in Taiwan is currently facing a tough competition. This study tries to provide an effective selection method for suppliers by combining the grey prediction, Data Envelopment Analysis (DEA), and heuristic technique. The model starts from DEA concept to find inputs and outputs, and then employs the grey prediction to predict the inputs and outputs, which are based on the historic data. The analysis includes DEA to measure the operation efficiency. One of the backlight module companies has chosen company A to be the target. Eight major listed TFT-LCD companies in Taiwan were selected as alternatives for partner selection. The empirical results show that some of eight companies who have better operation efficiency may result in poor efficiency afterwards. Contrarily, some companies with poor operation efficiency are having the chances of getting better. The proposed method can provide good advices for partners' selection.

Keywords: grey prediction; data envelopment analysis; alliance; supply chain

Applied System Innovation – Meen, Prior & Lam (Eds)
© 2016 Taylor & Francis Group, London, *ISBN 978-1-138-02893-7*

An investigation and policy suggestions on long-term unemployment in the Yunlin-Chiayi-Tainan Region

Chen-Yang Shih
Department of Technology Application and HRD, National Taiwan Normal University, Taiwan, R.O.C.

ABSTRACT: Long-term unemployment is one of the most difficult issues connected with employment. On the one hand, the development of the infrastructure of the workforce is seriously eroded because of prolonged unemployment. On the other hand, because the workers cannot return to the job market for a long time, some social problems may occur. Therefore, every country around the world keeps an eye on the situation of the long-term unemployed. In order to understand the long-term unemployment in the Yunlin-Chiayi-Tainan Region and to give suggestions on the services provided by the Regional Branch Workforce Development Agency, Yunlin-Chiayi-Tainan Region, and this research employed the questionnaire method, the interview method, and focused group discussions to collect first-hand data. The quantitative investigation was based on 837 long-term unemployed workers provided by the Regional Branch Workforce Development Agency, Yunlin-Chiayi-Tainan Region. 319 valid questionnaires were obtained. The researchers also recruited 30 interviewees who were selected from the long-term unemployment database. In addition, some experts were invited to participate in 3 focused group discussions to share their opinions on particular topics.

Applied System Innovation – Meen, Prior & Lam (Eds)
© 2016 Taylor & Francis Group, London, *ISBN 978-1-138-02893-7*

Understanding the users' willingness to help members in social media

Li-Wen Chuang & Shu-Ping Chiu
Department of Digital Media Arts and Design, Fuzhou University, Xiamen, Fujian, China

ABSTRACT: Given the rapid growth in the communication landscape for users brought about by participative Internet use, online games, and social media, it is important to explore a better understanding of these online technologies and their impact on the users' willingness to help a member. This study integrates perceived value, social capital, and cognitive absorption as the antecedents of the users' willingness to help individual members; furthermore affecting the Social Media Communities (SMCs) continuance. The results display the concernment of perceived value, social capital, and cognitive absorption, which plays a decisive role and produces direct effects in predicting the willingness to help online members in this model. Depending on our findings, practical implications for SMCs marketing strategies and theoretical implications will be predicted.

Other

Applied System Innovation – Meen, Prior & Lam (Eds)
© 2016 Taylor & Francis Group, London, ISBN 978-1-138-02893-7

The communication of sport facilities for caring for the aging society

M.T. Wang & C.T. Lin
Department of Industrial Design, National Kaohsiung Normal University, Kaohsiung, Taiwan

ABSTRACT: With the upcoming aging society, the concept of health care is being taken seriously. In addition to cultivate good dietary habits in individuals, regular exercise is an important way to achieve it. Each neighborhood park or the open space nearby has become the popular recreation and the sport areas for the aging group in Taiwan. It is urgently needed to communicate with the aging group about the healthy sports information. Recently, 16 popular kinds of outdoor fitness equipments have been added to these places. The willingness of the aging society to find for the adjusting equipment becomes worth exploring. We observed the neighborhood parks in Taiwan and surveyed 150 aging people over the age of 50 in a questionnaire, and we investigated their general sports habits and their willingness. We found that they concentrate on their sports time often in the early mornings and in the evenings, and they exercise with their family. The usability of the equipment is a maximum of 1–2 days a week, and within half an hour, it is found that the examinees do not frequently spend time on using the facilities. Moreover, they prefer lower limb movement equipment, such as the 'air-walker,' which is found to be the most popular one. They would like to use it because it makes them feel satisfied, and the main using purpose is for physical training. The related government division can communicate the results from this survey to develop or change more suitable equipment.

Applied System Innovation – Meen, Prior & Lam (Eds)
© 2016 Taylor & Francis Group, London, ISBN 978-1-138-02893-7

A study on the left-handed population's directional movement of the finger on touch devices

Tianyi Qin
College of Physics and Information Engineering, Fuzhou University, Fuzhou, Fujian, P.R. China

Jun He
Xiamen Academy of Arts and Design, Fuzhou University, Xiamen, Fujian, P.R. China

ABSTRACT: The directional movement of users' fingers has become one of the most interesting topics for human-computer interaction. Since a large proportion of people are right-handed, most application software for touch screen devices are designed based on the habit of right-handed users. However, this kind of design may cause inconveniences for the left-handed users when they slide over the screen, and it might lead to operational mistakes. In this study, a tunnel sliding experiment is designed to investigate the finger behavior attributes of the left-handed population. According to the experimental results, the finger behavioral differences between the left- and right-handed people are figured out, which contributes to the design of the user friendly software of touch screen devices for both the left- and right-handed users. This study is the basic research of touch devices interaction design and has a guiding value for the user interface design of mobile the touch devices.

Applied System Innovation – Meen, Prior & Lam (Eds)
© 2016 Taylor & Francis Group, London, ISBN 978-1-138-02893-7

Application of mirror theory in fire-fighting equipment design

Ding-Bang Luh & Ottavia Huang
National Cheng Kung University, Tainan, Taiwan, R.O.C.

ABSTRACT: This study introduces a cognitive design framework that emphasizes on 'positive creativity' to assist designers in creative process under the notion of creation as "bringing something out of nothing", switching one's perception of problems as a potential source of creative ideas which will satisfy both producers and customers. It brings designers to think beyond the scope of problems (in producer's perspective) and expectations (in customer's perspective) under the framework of the Mirror Theory, which transforms problems/complaints into expectations and redefine design as an action to fulfill wishes that satisfy both idea adopters (producers and costumers) and end-users. The application of Mirror Theory has been tested on more than 20 design cases, ranging from creations of products, services, system to formulation of strategy, branding, and marketing. This study applied the Mirror theory in the design of fire-fighting equipment that increases firemen's safety, line of visions, and mobility during their duty.

Applied System Innovation – Meen, Prior & Lam (Eds)
© 2016 Taylor & Francis Group, London, ISBN 978-1-138-02893-7

Experimental teaching research on spaciousness of sculpture and deepness of form based on "Empirical-Interaction"

Li Ma & Qian Jiang
Xiangtan University, China

ABSTRACT: This research mainly guides students to perceive the relationship between spaciousness and deepness of form during the process of sculpture creation through numerous approaches of "Empirical-Interaction," in order to complete the interpretation of original purpose from artworks. This research establishes the logical system of "Empirical-Interaction," leading the students to utilizing this logical system to think deeply and finally creating homologous artworks. The course comprises three modules: creation of original artworks, interposition of "Empirical-Interaction" teaching pattern, and recreation of artworks. On the theoretical basis, the "cone of experience" introduced by Dale involves the digital sculpture and makes students achieve a new perception to the judgment of spaciousness and create a sculpture with an even more profound expression. Based on the experimental process of leading the students to constructing the "Empirical-Interaction," the cognitive transformation of the deepness of physical volume will be embodied in students' sculptures.

Keywords: instructional design; "Empirical-Interaction"; spaciousness of sculpture; deepness of form

Applied System Innovation – Meen, Prior & Lam (Eds)
© 2016 Taylor & Francis Group, London, *ISBN 978-1-138-02893-7*

Practical research on "performance" teaching model in the *Sound Design* course of animation majors in colleges and universities

Qian Jiang & Li Ma
Xiangtan University, China

ABSTRACT: With the growing establishment of the *Sound Design* course of animation majors in colleges and universities, under the new situation, the content and teaching model of the *Sound Design* course has been demanded to be more professional to meet the requirements of the animation creation and industry market. The introduction of the "Performance" teaching model in the *Sound Design* course, which largely stimulates creative thinking in animation, is of significant importance in motivating the students to create animation works with initiative. In this teaching model, animation ideas and thinking skills will interact and improve under the guidance of sound theory, psychology, and performance, making the animation idea thinking modes even more diversified. Furthermore, the teaching model attempts to influence the teaching pattern of other courses in animation and strengthen the integration of the major, and makes it more adapted to the needs of the current market for college talents.

Keywords: "performance" teaching method; sound design; animation course; creative thinking of animation

Applied System Innovation – Meen, Prior & Lam (Eds)
© 2016 Taylor & Francis Group, London, *ISBN 978-1-138-02893-7*

The study on the eye tracking data collected from watching animated movies

Li Qing & Miao Li
Department of Visual Communication, Art Institute of Xiangtan University, Xiangtan, P.R. China

ABSTRACT: The acquirement of target viewers' feedback on an animated movie, plays a critical role in the movie's post-production as well as its distribution strategy into market. The traditional acquisition methods, such as surveys, test screenings, and interviews, could be easily impacted by the viewers' social desirability effect and recall errors. In this study, the team applied eye tracker, which has been used widely in advertising and marketing, to collect viewer's visual responses to selected animation clips. The raw data collected by eye tracker were summarized into 4 key measurements we defined to suit the animation product's unique features. The defined measurements are eye wandering rate, character attention density, visual guidance effect, and visual resistance.

Keywords: animation; visualization; eye tracker

Applied System Innovation – Meen, Prior & Lam (Eds)
© 2016 Taylor & Francis Group, London, ISBN 978-1-138-02893-7

The study on VJing—a new combination of visual art with life visual performance

Gao Shen
Department of Visual Communication, Art Institute of Xiangtan University, Xiangtan, P.R. China

ABSTRACT: VJing, a short form of "Video Jockeying", is a new kind of media art, which has a short history but large influence in visual arts. It is a manifestation which has come to eminence and become a kind of prevalent art, music, and performance during the last 10 years. VJing is normally misinterpreted as a live form of MTV. However, VJing is actually to see how designers changing the rhythm, hue, or shape of the images in a live visual performance. Since people have no proper principle to evaluate the VJing works, artists are explaining and presenting the VJing works in their own ways. In addition, the relationship between screen and image is based on the connection of projector installation, so the designer and his/her crew have to do many hard works before the performance. The whole paper introduces VJing theory and elements, compares VJing with film, interprets VJing environment and audience and analyzes several VJing examples. According to the paper, VJing can be the media that connects these two aspects and makes a global party that will last permanently.

Keywords: VJing; music; live visual performance; audience; image

Applied System Innovation – Meen, Prior & Lam (Eds)
© 2016 Taylor & Francis Group, London, ISBN 978-1-138-02893-7

Study on the mode of representation of the traditional grass cloth weaving techniques from the perspective of 'Interactive Design'

Zhun Huang
Xiangtan University, Xiangtan, Hunan, P.R. China

ABSTRACT: To preserve the grass cloth-weaving techniques, recording and profiling are often employed to present the historical documents and relics about the traditional artistry. To promote it, innovating the technical procedures or reinventing the applications of the product (Chinese linen made of grass cloth) are usually adopted. However, the one-way recording, profiling or applying fails to deliver the touch, the texture, the charm and the whole experience to the audience; because the soul of grass cloth-weaving hides in the intricate and magic hand-making process. This paper has borrowed the information interaction theory to illustrate the traditional grass cloth-weaving techniques and to systematically analyze the following five aspects, namely, "feel", "action", "interpersonal", "tools" and "Scene". It intends to shed light on the nature of grass cloth-weaving and render the audience an alternative viewing perspective and a more touching, unforgettable and thought-provoking experience.

Applied System Innovation – Meen, Prior & Lam (Eds)
© 2016 Taylor & Francis Group, London, ISBN 978-1-138-02893-7

Study on the explosive crack in the organic glass of a trainer aircraft cockpit

B.L. Dong, J.J. He, W.F. Zhang, W.T. Lou & H.P. Chen

School of Reliability and System Engineering Science and Technology Laboratory on Reliability and Environment, Beihang University, Beijing, China

ABSTRACT: After the trainer aircraft was serviced for a period, an explosive crack was found in the bonding position between the transparent parts and the polyester ribbon. The YB-3 organic glass was colorless and transparent. The intensity of the YB-3 organic glass was slightly higher than that of the YB-2 organic glass. Its thermostability was also higher than that of the YB-2. The YB-3 organic glass had good aging resistance and processing performance. The features and forming mechanism of explosive cracks in the YB-3 organic glass were investigated. Low-fold morphology video microscope was employed to observe the macromorphology. Environmental scanning electron microscope (ESEM) was employed to analyze the fracture morphology. Based on these observations and analyses, it can be concluded that the explosive crack is a fatigue crack and its formation is independent of the process defects. In addition, the craze at the tip of the explosive crack starts to initiate after the explosive crack is formed.

Keywords: explosive crack; low-fold morphology; ESEM; craze

Applied System Innovation – Meen, Prior & Lam (Eds)
© 2016 Taylor & Francis Group, London, *ISBN 978-1-138-02893-7*

Fluid dynamic characteristic analysis of an eccentric butterfly valve

T.J. Wu
Institute of Nuclear Energy Research Atomic Energy Council, Taiwan

C.K. Fang & G.C. Tsai
Department of Mechanical and Electro-Mechanical Engineering, National Ilan University, Yilan, Taiwan

ABSTRACT: In this research, Computational Fluid Dynamic (CFD) software was used in the simulation analysis of an eccentric butterfly valve. The opening angle of the eccentric butterfly valve increased from 0°to 90° (fully opened), with a 10% increase in each step. The flow Coefficient (CV values) was obtained in each step. The pressure, flow velocities, and turbulence distribution in each case were observed in the post-processing. In the simulation, two different turbulence models, K–ε and Reynolds Stress Method (RSM), were used and their applicability to valves with different opening angles was investigated. The results showed that the opening angle from 40% to 60% would be of better use. Both the RSM turbulence model and the K–ε turbulence model were suitable for the opening of 70% to 100%. The CV values were found to be proportional to the opening degree. The percentage errors between analytical and experimental CV values were within 5% except at singular points.

Keywords: eccentric butterfly valve; computational fluid dynamic analysis; flow Coefficient (CV); turbulence model; singular points

Applied System Innovation – Meen, Prior & Lam (Eds)
© 2016 Taylor & Francis Group, London, *ISBN 978-1-138-02893-7*

Application of robust design to a process simulation of injection molding automotive parts

W.T. Huang, D.H. Wu & C.L. Tsai
National Pingtung University of Science and Technology, Neipu, Pingtung, Taiwan

ABSTRACT: The objective of this study is to use computer-aided engineering to analyze injection molding of automotive parts using a robust design method. In the original process, the mold often filled incompletely the "short shot" phenomenon. Analyzing the process parameters, it was determined that excessive shear stresses in the plastic were the cause. Employing robust design optimization, the adjusted control parameters were plastic material temperature, mold temperature, injection pressure and injection time. This study showed that the factor with maximum influence on shear stress is the plastic material temperature. The optimization of injection parameters using robust design can significantly reduce the shear stress in plastics, and effectively mitigate short shot.

Keywords: robust design; short shot; automotive parts; shear stress

Applied System Innovation – Meen, Prior & Lam (Eds)
© 2016 Taylor & Francis Group, London, ISBN 978-1-138-02893-7

The application of Support Vector Machine approach in studying cardiotoxicity of Traditional Chinese Medicine compounds

J.F. Zhang, L.D. Jiang & Y.L. Zhang
School of Chinese Material Medica, Beijing University of Chinese Medicine, Beijing, China

ABSTRACT: In this study, 71 compounds with cardiotoxicity and 87 compounds with no cardiotoxicity were selected as a data set for the construction of cardiotoxicity discriminative model (model I). In addition, hERG potassium ion channel blocker discriminative model (model II) was constructed based on 45 blockers and 49 non-blockers. Two models with high accuracy were built by using the Support Vector Machine (SVM) approach. The accuracy, sensitivity, specificity and Matthews's correlation coefficient of the two models were all above 85%. Model I was further evaluated based on 22 cardiotoxic traditional Chinese medicine compounds. Then, by using model II, 5 potential compounds, which may lead to cardiotoxicity through blocking hERG potassium ion channel, were screened from the positive results of model I. Then, two discriminative models were utilized to identify the potential cardiotoxic compounds from Traditional Chinese Medicine Database (TCMD) and analyzed the mechanism of them.

Keywords: support vector machine, traditional Chinese medicine, cardiotoxicity, hERG potassium ion channel blocker

Applied System Innovation – Meen, Prior & Lam (Eds)
© 2016 Taylor & Francis Group, London, *ISBN 978-1-138-02893-7*

Comparative analysis of boron steel B1500HS constitutive models

Q.L. Wang, B.T. Tang & W. Zheng
Institute of Engineering Mechanics, Shandong Jianzhu University, Jinan, China

ABSTRACT: In this paper, the isothermal uniaxial tensile tests for boron steel B1500HS were performed by a Gleeble-1500D thermo-mechanical simulator. The true stress-strain data were obtained under different loading conditions. The flow behavior of boron steel B1500HS was described using the Hensel–Spittle model, Nemat–Nasser model and Tong–Wahlen model, respectively. The predictions of these constitutive models are compared with each other using statistical measures like correlation coefficient, average absolute relative error and its root mean square error. The results show that the Hensel-Spittle model seems to have higher accuracy and reliability in predicting the flow behavior.

Keywords: boron steel; constitutive equation; flow stress; tensile test

Modeling and Simulation of Mechanical Systems

Applied System Innovation – Meen, Prior & Lam (Eds)
© 2016 Taylor & Francis Group, London, ISBN 978-1-138-02893-7

On the effect of blade twist angle on Horizontal-Axis Wind Turbine performance

Huei Chu Weng & Hui-Ren Huang
Department of Mechanical Engineering, Chung Yuan Christian University, Taoyuan, Taiwan

ABSTRACT: It is desirable to understand the role of blade twist angle in wind turbine performance analysis. In this study, the influence of flake-type blade twist angle on the performance of a horizontal-axis wind turbine for different wind speeds and blade numbers is investigated. The wind tunnel experiments are performed in a low-speed, open circuit wind tunnel, and the CFD software STAR CCM+ is used to provide valuable information for the understanding of flow behavior. Results reveals that, when the blade twist angle increases, the power output and conversion efficiency increase to their maxima and then decrease for different wind speeds and blade numbers. It is also found that increasing the wind speed and blade number can further magnified the critical twist angle.

Applied System Innovation – Meen, Prior & Lam (Eds)
© 2016 Taylor & Francis Group, London, ISBN 978-1-138-02893-7

The Magnetohydrodynamic field on the surface instability of axisymmetric condensate films at liquid–vapor interfaces

Po-Jen Cheng & Kuo-Chi Liu
Department of Mechanical Engineering, Far-East University, Tainan, Taiwan

David D.W. Lin
Graduate Institute of Mechatronic System Engineering, National University of Tainan, Tainan, Taiwan

ABSTRACT: It is highly desirable that higher performances for homogeneous film growth can be developed in various industrial applications. Magnetohydrodynamics (MHD) is the study of the interaction of electrically conducting fluids and electromagnetic forces. In magnetohydrodynamic flows, the conducting magnetic field and the fluid strongly interact and create complex magnetic and dynamic phenomena. In practice, the optimum conditions can be found through the use of a system to alter the stability of the film flow by an appropriate control of the applied magnetic field. Flow governing the equation on thin film flow is solved on an order-by-order basis by the perturbation method. The multiple scale method is used to derive the Ginzburg Landau equation, and a filtered wave with no spatial modulation is considered to analyze the flow criteria and classify the regions where the threshold amplitude exists. The modeling results indicate that the degree of the stability has been increased in the film system, with a condensate effect. Furthermore, it is shown that by increasing the magnetic field and cylinder radius its stability tends to get enhanced as it travels down toward the vertical cylinder.

Applied System Innovation – Meen, Prior & Lam (Eds)
© 2016 Taylor & Francis Group, London, ISBN 978-1-138-02893-7

Vibration control of a structure using a shape memory material absorber based on fuzzy system

Chih-Jer Lin
Graduate Institute of Automation Technology, National Taipei University of Technology, Taipei, Taiwan

Chun-Ying Lee
Department of Mechanical Engineering, National Taipei University of Technology, Taipei, Taiwan

Ting-Yu Chen
Graduate Institute of Automation Technology, National Taipei University of Technology, Taipei, Taiwan

ABSTRACT: For on-line process of semiconductor, the vibration control is very important to perform precise measurements or increase the production yield rate. To improve vibration control for industrial applications, this study is to develop a real-time fuzzy semi-active controller for a tunable vibration Absorber (TVA) which is made of Hybrid Sharp Memory Materials (HSMM). Because the characteristic of the SMM is dominated by the temperature, the fuzzy control is used to capture the Vibration-Temperature mapping based on real-time vibration experiments. To improve the performance of the HSMM TVA, a fuzzy controller is proposed and the adaptive semi-active controller is used to suppress the vibration of a plane structure.

Applied System Innovation – Meen, Prior & Lam (Eds)
© 2016 Taylor & Francis Group, London, *ISBN 978-1-138-02893-7*

Adaptive control design for Induction Motor with uncertain parameters: An LMI approach

Ming-Fa Tsai & Chung-Shi Tseng
Department of Electrical Engineering, Minghsin University of Science and Technology, Hsinchu, Taiwan

Eric Chen
Rich Electric Co. Ltd., Tainan, Taiwan

ABSTRACT: In this paper, an adaptive control scheme is proposed for the Induction Motor (IM) with uncertain parameters using an LMI approach. A novel formulation for the uncertain IM motor drive system is developed. By the formulation, the stability condition for the closed-loop system with the proposed adaptive controllers can be characterized in terms of some Linear Matrix Inequalities (LMIs). Particularly, the speed controller and the current controller are considered simultaneously and the control gains in both the speed loop and the current loop can be obtained systematically. Numerical solution for the control gains and simulation results using MATLAB LMIP toolbox and MATLAB/Simulink tool are provided to illustrate the design procedure and the performances.

Applied System Innovation – Meen, Prior & Lam (Eds)
© 2016 Taylor & Francis Group, London, ISBN 978-1-138-02893-7

Analysis of heat transfer in a pulse laser-irradiated tissue

K.C. Liu
Department of Mechanical Engineering, Far East University, Tainan, Taiwan

F.J. Tu
Department of Electrical Engineering, Nan Jeon University, Tainan, Taiwan

P.J. Cheng
Department of Mechanical Engineering, Far East University, Tainan, Taiwan

ABSTRACT: A generalized dual-phase-lag bioheat transfer equation was derived from a two-temperature model. This paper uses it to describe the heat transfer in a pulse laser-irradiated tissue. There are mathematical difficulties for solving such a problem. Therefore, a suitable and efficient method is required for the accurate and stable solutions. The problem is solved with the hybrid application of the Laplace transform and the modified discretization technique. The paper notices that the phase lag times depend on the porosity, heat capacities of blood and tissues, coupling factor, and the ratio of thermal conductivity of tissue and blood. A discussion to the discrepancy of the present results with those in the literature is made. Results show the generalized dual-phase-lag bioheat transfer equation does not reduce to the Pennes bio-heat transfer equation for $\tau_q = \tau_T$.

Applied System Innovation – Meen, Prior & Lam (Eds)
© 2016 Taylor & Francis Group, London, ISBN 978-1-138-02893-7

Error modeling for the leg mechanism of the quadruped walking chair robot

H.B. Wang
Parallel Robot and Mechatronic System Laboratory of Hebei Province, Yanshan University, Qinhuangdao, HeBei Province, China
Key Laboratory of Advanced Forging and Stamping Technology and Science, Ministry of Education, Yanshan University, Qinhuangdao, HeBei Province, China

S.S. Wang, Y.H. Wen & N. Chen
Parallel Robot and Mechatronic System Laboratory of Hebei Province, Yanshan University, Qinhuangdao, HeBei Province, China

ABSTRACT: To reduce the output error of the parallel robot, it is crucial to analyze the pose error of the parallel mechanism. This paper deals with the pose error analysis, and establishes the model for the effects of pose parameters and structural parameters on the accuracy of the leg mechanism. Firstly, based on the research of constraint relations on the leg mechanism of the walking chair robot, the error model is established. Then the influence of structure parameters and pose parameters on the pose error is studied. The results contribute a lot to the determination of the structural parameters and the output error of the control system.

Applied System Innovation – Meen, Prior & Lam (Eds)
© 2016 Taylor & Francis Group, London, ISBN 978-1-138-02893-7

Optimal manufacturing process of the mulberry on high GABA metabolites

Y.L. Yeh, M.J. Jang & H.J. Sheu
Department of Automation and Control Engineering, Far East University, Taiwan

Z.S. Cai
Department of Computer Application Engineering, Far East University, Taiwan

ABSTRACT: This paper studies the effect of an optimal manufacturing process of the mulberry on high GABA metabolites. These manufacturing parameters include the ultrasonic vibration time, withering time, aerobic fermentation time, and anerobic fermentation times. The Taguchi Methods are applied to get the optimal process on the high GABA metabolites. From the analyses results, it can be known that the effect of the ultrasonic vibration time and anerobic fermentation for the mulberry on GABA metabolites is quite large. As the Taguchi method is applied to get high GABA metabolites, this can get these optimal parameters. When these optimal parameters are applied to the mulberry tea manufacturing process, we can then get high GABA metabolites (482.02 mg/100 g). Therefore, it can be known that the optimal manufacturing process can raise the GABA metabolites for the mulberry.

Applied System Innovation – Meen, Prior & Lam (Eds)
© 2016 Taylor & Francis Group, London, *ISBN 978-1-138-02893-7*

Analysis of driver's musculoskeletal characteristics and vehicle handling properties under different power-assisted steering systems

Zhenhai Gao & Da Fan
State Key Laboratory of Automobile Simulation and Control, Jilin University, Changchun, China

Kaishu Zhao
The First Affiliated Hospital of Jilin University, Changchun, China

Hui Zhao & Huili Yu
State Key Laboratory of Vehicle NVH and Safety Technology, Changan Automobile Holding Ltd., Chongqing, China
Changan Automobile Holding Ltd., Automotive Engineering Institute, Chongqing, China

ABSTRACT: Based on the musculoskeletal model, a driver's steering maneuver of driving vehicles equipped with different steering systems was simulated using inverse dynamics. Key vehicle handling metrics and the driver's physiological parameters were extracted. Multiple linear regression was therewith established between them, using the driver's physiological parameters as dependent variables. The regression models showed a quantitative relationship between certain vehicle handling metrics and the driver's physiological parameters, and validated their significant correlation. Vehicle handling metrics that significantly affect the driver's physiological reactions were discussed. These results can serve as references for a steering performance evaluation using the driver's physiological characteristics.

Applied System Innovation – Meen, Prior & Lam (Eds)
© 2016 Taylor & Francis Group, London, ISBN 978-1-138-02893-7

Optimal hanger locations for automotive exhaust systems

W.Z. Du
School of Mechatronic Engineering Technology, Xuzhou College of Industrial Technology, Xuzhou, China

K.N. Chen, W.H. Liu & S.J. Chen
Department of Mechanical Engineering, Tungnan University, New Taipei City, Taiwan

Y.L. Hwang
Department of Mechanical Design Engineering, National Formosa University, Yunlin County, Taiwan

ABSTRACT: Excited by the exhaust gas pulsation and the vibrations from the engine and other components, the exhaust system of an automobile significantly influences the vehicle's NVH performance. In this research, the hanger locations for an automotive exhaust system are optimized to reduce the possibility of resonance between the exhaust system and the car body. Finite element analysis on the exhaust system is first performed by modeling the stiffness coefficients of the hanger isolators as sets of translational springs, and then the hangers are relocated to optimal positions such that the natural frequencies of the new system are separated from the idling speed of the running engine. The optimum design is a feasible design with all its FEA frequencies being outside the danger zone, and it should enhance the vehicle's NVH performance and increase the durability of the exhaust system.

Applied System Innovation – Meen, Prior & Lam (Eds)
© 2016 Taylor & Francis Group, London, ISBN 978-1-138-02893-7

Nonlinear vibration active control of structure

Shueei-Muh Lin & Teng Minjun
Mechanical Engineering Department, Kun Shan University, Tainan, Taiwan, R.O.C.

ABSTRACT: In this study, the nonlinear vibration model of structure with cross support is established. The conventional structure without cross support is linear and easy to be investigated. Unfortunately, its dynamic stability and vibration due to earthquake excitation are usually not acceptable. For suppressing the structural vibration the cross support composed of the elastic connecting bar and damper is considered here. This is a passive control design. Beside, due to the supporting arrangement, ,the mathematical model of the structure is highly nonlinear. In this study, the analytical solution for this system is derived. Moreover, the effect of the cross support on suppression of structure vibration is effective but limited. For increasing the effectiveness of vibration control, the active-tuned-mass system is designed. Further, the effects of control parameters on the vibration response are investigated. It is found that under the synchronous control the vibration response ratio of the second floor approaches to zero.

Keywords: nonlinear vibration; structure; analytical solution; synchronous control

Automation and Intelligent Systems

Applied System Innovation – Meen, Prior & Lam (Eds)
© 2016 Taylor & Francis Group, London, ISBN 978-1-138-02893-7

Social media: Explore the most influenced determinants of the social media to enhance customer's satisfaction in current hyper-competitive M-commerce

Ming-Yuan Hsieh
Department of International Business, National Taichung University of Education, Taichung, Taiwan

Chuen-Jiuan Jane
Department of Finance and Risk Management, Ling-Tung University, Taichung, Taiwan

ABSTRACT: Beyond the potential for significant profits and developing a niche in the current hyper-competitive m-commerce era, a majority of Taiwanese enterprises has devoted themselves to discover the most determinants of social media to achieve the highest customer's satisfaction for the highest corporate profits. Hence, this research not only orderly employs the House of Quality (HOQ) model of the Quality Function Development (QFD) method and Analytical Network Process (ANP) to induce the correlationship between social media and customer satisfaction by exploring the most influenced determinants of social media, but also further applies a diversified comparison of the Multiple Criteria Decision Making (MCDM) methodology in statistic comparable measurements for the increment of research reliability and validity. Deductively, in accordance with a series of accurately measured consequences, it is a very valuable contribution that Taiwanese companies have to develop the most critical and influenced determinants of the social media: the User Generated Contents (UGC), the Behavior Targeting (BT) and the Rich Media (RM) achieve the extensive customers' desires in order to create the most beneficial and optimal profit margins.

Applied System Innovation – Meen, Prior & Lam (Eds)
© 2016 Taylor & Francis Group, London, *ISBN 978-1-138-02893-7*

A novel approach of botnet feature analysis

S.L. Chen, Y.Y. Chen, Y.T. Cheng, S.H. Kuo & H.P. Wang
Institute of Manufacturing Information and Systems, National Cheng Kung University, Tainan City, Taiwan, R.O.C.

ABSTRACT: Computers infected by botnets can be remotely controlled by attackers on the cloud, which is one of the most common attack platforms on software systems and has become an important issue in the information security domain. In this paper, we collect and analyze the suspicious cloud network flow information by setting up experiment environments for observing different infected machines for extracting features of real-world botnets and monitoring the machines connected to botnet black lists in the network of National Cheng Kung University.

Keywords: Malware; information security; botnet feature ontology; cloud computing

Applied System Innovation – Meen, Prior & Lam (Eds)
© 2016 Taylor & Francis Group, London, ISBN 978-1-138-02893-7

iAgentX: A novel cloud multi-agent architecture for machine information management

Shang-Liang Chen, You-Chen Lin & Yun-Yao Chen
Institute of Manufacturing Information and Systems, National Cheng Kung University, Tainan City, Taiwan, R.O.C.

ABSTRACT: In this paper, we proposed a novel cloud multi-agent architecture for machine information management based on an all-electric injection molding machine and its' information retrieval agent development via a signal retrieval technology called iAgentX. This architecture can be used for production information integration from a machine into a cloud data center for machine information management through cloud agent off-line and non-invasive technology. Windows Management Instrumentation (WMI) technology is utilized to construct our approach framework and to serve as the multi-agent cloud management technology.

Applied System Innovation – Meen, Prior & Lam (Eds)
© 2016 Taylor & Francis Group, London, ISBN 978-1-138-02893-7

Improved customer satisfaction-based novel decision support system on airlines service by using the Particle Swarm Optimization method

Chung-Lin Huang
Department of Tourism Management, Taiwan Shoufu University, Tainan City, Taiwan, R.O.C.

Chung-Chi Huang
Department of Automation and Control Engineering, Far East University, Tainan City, Taiwan, R.O.C.

ABSTRACT: The main purpose of this research is to apply a Novel Decision Support System to improve the customer satisfaction of airlines service. In the past, the measurement of customer satisfaction was made by using a questionnaire. We collect and compute the customer satisfaction of each flight by the cloud computing-based decision support system. It compares the customer questionnaire, and then continuously modifies the accuracy of the prediction system by using the Particle Swarm Optimization (PSO) method. In previous research, most of the DSS were focused on the analysis process. It this study, we develop a system that provides a precise solution instead of analysis. It efficiently and precisely improves the customer satisfaction for the long-term perspective. We propose a parameter of evaluation module that selects 12 influence factors from MEPH and 12 satisfaction evaluation factors of airlines service (by customer perception of service quality). In this study, we use the Particle Swarm Optimization (PSO) method to build the module. The module can be used by calculating the airlines service satisfaction from the quality factors such as material, machine, product, and staff. However, we can obtain the prediction of service satisfaction by means of data fusion by using the 12 satisfaction indicators. The results can be used to make the service quality strategy, in order to lead to a higher customer satisfaction. These findings can help airlines managers to predict their customer satisfaction more efficiently, making changes to the service quality strategy easily to meet the customers' satisfaction level. Even if the managers pre-set a customer's satisfaction level, a real-time cloud computing can help managers deploy the resources to achieve the goal. Finally, in order to prove the feasibility of the parameters and the intelligent evaluation methodology, this study collects the data and tests the evaluation of quality from an experiment conducted on an airlines service system in Kaohsiung City in Taiwan.

Applied System Innovation – Meen, Prior & Lam (Eds)
*© 2016 Taylor & Francis Group, London,**ISBN 978-1-138-02893-7*

Omnidirectional spherical robot

Chih-Hui Chiu & Yao-Ting Hung
Department of Electrical Engineering, Yuan-Ze University, Tao-Yuan, Taiwan, R.O.C.

ABSTRACT: In this study, a Supervisory Tracking Control System (STCS) is proposed for an Omnidirectional Spherical Robot (ODSR) control problem. The stability of the STCS, which is based on the Lyapunov stability theorem, can be ensured without any strict constraint. Finally, the efficiency of the supervisory tracking control system is verified by ODSR real-world implementation.

Applied System Innovation – Meen, Prior & Lam (Eds)
© 2016 Taylor & Francis Group, London, *ISBN 978-1-138-02893-7*

A defect type recognition system for underground power cable joint based on Semi-Supervised Learning algorithm

Horng-Lin Shieh, Cheng-Chien Kuo & Chi-Chang Huang
St. John's University, New Taipei City, Taiwan, R.O.C.

ABSTRACT: This paper proposed a hybrid learning algorithm for defect type recognition system for a 25kV *Cross-Linked Polyethylene* (XLPE) underground power cable joint. In data analysis method, the Semi-Supervised Learning (SSL) is adopted when parts of the data are unlabeled. The primary concept of SSL involves labeled data being used to evaluate the class of unlabeled data. In this paper, a SSL which integrated the fuzzy-rough set and *Shared Nearest Neighbors* (SNN) is proposed. A defect type recognition system XLPE cable classification problem is used in order to test the proposed algorithm. The experimental results show that the proposed algorithm can obtain outstanding levels of performance.

Applied System Innovation – Meen, Prior & Lam (Eds)
© 2016 Taylor & Francis Group, London, ISBN 978-1-138-02893-7

Power cable insulation defect identification using ANN-based pattern recognition approach

Po-Hung Chen, Li-Ming Chen, Cheng-Chuan Chen & Han-Chien Chen
Department of Electrical Engineering, St. John's University, New Taipei City, Taiwan

Ming-Chang Tsai
Department of Electronic Engineering, St. John's University, New Taipei City, Taiwan

ABSTRACT: This paper presents an ANN-based pattern recognition approach for identifying the insulation defects of power cables. More than half of breakdown accidents of power cables were caused by insulation deterioration. Partial discharge pattern recognition has been regarded as an important diagnosis method to prevent power cables from malfunction of insulation defect. In this work, four 25 kV power cables with typical insulation defects are purposely made. These power cables will be used as the experimental models of partial discharge examination. A precious partial discharge instrument is used to measure the 3-D partial discharge signals of these power cables in a shielded laboratory. The databases of 3-D partial discharge signals are used as the training data to train a three-layer ANN. The training-accomplished ANN can be a good insulation defects identification system for power cables. The proposed approach is successfully applied to practical power cables field experiments. The result of experiments shows that the recognition rates of the proposed approach are superior to the existing methods in literatures.

Applied System Innovation – Meen, Prior & Lam (Eds)
© 2016 Taylor & Francis Group, London, ISBN 978-1-138-02893-7

Forecasting house prices using DCSVR

Chen-Yuan Lee & Jui-Chung Hung
Department of Computer Science, University of Taipei, Taipei, Taiwan

ABSTRACT: This paper reports forecasting house prices by using a District Classification Support Vector Regression (DCSVR) model. In general, house prices have nonlinear and clustering properties. Thus, in this study, a Support Vector Regression (SVR) model was used to perform district classification and the results are presented in this paper. We executed the following procedures: First, the house data were classified according to the district in which the house is located. Second, the classifications conducted in the first stage were used to establish the SVR model, which was used to reduce the complexity of the problem.

The proposed model is highly nonlinear and complex. We present an iterative algorithm based on a genetic algorithm for optimizing the parameters of the DCSVR model. Data of actual prices of registered houses in Taipei, Taiwan, from November 2013 to December 2014 were used to demonstrate the effectiveness of the proposed model. The simulation results indicated that the model performance improved considerably regarding house price prediction; the results also indicated that the accuracy of the proposed model was higher than that of the SVR model.

Applied System Innovation – Meen, Prior & Lam (Eds)
© 2016 Taylor & Francis Group, London, *ISBN 978-1-138-02893-7*

Intelligent assessment and prediction for a customized physical fitness and healthcare system

Chung-Chi Huang
Department of Automation and Control Engineering, Far East University, Tainan City, Taiwan, R.O.C.

Hsiao-Man Liu
Department of Recreational Sports Management, Far East University, Tainan City, Taiwan, R.O.C.

Chung-Lin Huang
Department of Tourism Management, Taiwan Shoufu University, Tainan City, Taiwan, R.O.C.

Yen-Ting Ke
Department of Mechanical Engineering, Far East University, Tainan City, Taiwan, R.O.C.

ABSTRACT: In the advent of global high-tech industry and commerce era, sedentary life reduces opportunities of physical activity. Thus, the physical fitness and health of people is getting worse day by day. Therefore, it is necessary to develop a system that can enhance physical fitness and health for people. But it is hard to meet the needs of an individual in a general physical fitness and healthcare system. The main purpose of this research is to develop an intelligent assessment and prediction for a customized physical fitness and healthcare system. It records all processes of the physical fitness and healthcare system by wireless sensors network. The results of the assessment and prediction for a customized physical fitness and healthcare system will be generated by the inference of a fuzzy logic and neural network. It will improve an individual's physical fitness and healthcare. Finally, we will demonstrate the advantages of intelligent assessment and prediction for a customized physical fitness and healthcare system.

Applied System Innovation – Meen, Prior & Lam (Eds)
© 2016 Taylor & Francis Group, London, ISBN 978-1-138-02893-7

Sliding mode flight control design for an aircraft

Kuan-Chen Lin, Shyi-Kae Yang & Tzu-Yin Chang
Department of Automation and Control Engineering, Far East University, Tainan, Taiwan

ABSTRACT: Flight of an aircraft is a motion with six Degrees-of-Freedom (DOF) that is depicted by highly coupled and nonlinear dynamics. The task of proper controller design for aircraft flight control mission still remains a challenging problem in the literature of flight automation studies. For nonlinear system controls, a technique called Sliding Mode Control (SMC) is a very powerful approach; additionally, system robustness is also a benefit. It can be applied in various complex system platforms, such as ships, helicopters, tanks and of course, aircrafts. However, traditional SMC methods, often require full measurement of system states and infinite actuator bandwidth; otherwise a problem caused by infinitely high-frequency actuator switching called chattering phenomenon will occur. These two major drawbacks confine the realization of the traditional SMC methods in real world applications. Therefore, this paper proposes a new sliding mode control approach and applies it to the problem of flight control design for an aircraft platform. In the proposed approach, the output feedback is used to control the flight dynamics such that the aircraft can be piloted automatically. The synthesis of the control utilizes only output feedback information to operate the flight of an aircraft. The control is well-designed and is very similar to the real -life flight situation but free from the actuator chattering effects. The results suggest that the entire proposed control approach is effective and can be very helpful for practical realization of real aircraft flight.

Applied System Innovation – Meen, Prior & Lam (Eds)
© 2016 Taylor & Francis Group, London, *ISBN 978-1-138-02893-7*

Development of visual control interface for a mobile robot with a single camera

Chun-Tang Chao, Ming-Hsuan Chung, Chi-Jo Wang & Juing-Shian Chiou
Department of Electrical Engineering, Southern Taiwan University of Science and Technology, Tainan, Taiwan

ABSTRACT: This paper proposes the development of a visual control interface for a mobile robot with a single camera. The mobile robot, equipped with a single camera, can be remote-controlled by a smartphone with Android system, and the real-time video is transmitted based on WiFi. The robot employs the Arduino Yun development board as its core and the Android app is developed in the controller end. With the video transmission function, the robot can execute the task out of the controller's sight. This makes the robot much more functional. Moreover, the robot has a three-axis robotic arm to finish the robotic grasping. But in practical application, it usually takes more time for the user to manipulate the robotic arm to the desired position, especially in an emergency situation. By clicking on the desired position or object on the screen, the proposed robot will calculate the target's 3D location by motion stereo vision and formulate the robotic arm to reach the target position autonomously. If the target is in the working space of the robotic arm, the robotic arm will be driven to approach the approximate target position, and then the exact position can be reached by the user in a quicker way.

Applied System Innovation – Meen, Prior & Lam (Eds)
© 2016 Taylor & Francis Group, London, ISBN 978-1-138-02893-7

Design of output feedback controllers for discrete Takagi-Sugeno fuzzy systems

K. Hoshino & J. Yoneyama
Department of Electronics and Electrical Engineering, Aoyama Gakuin University, Tokyo, Japan

ABSTRACT: This study deals with the design of static output feedback controllers for discrete-time Takagi-Sugeno fuzzy systems. We present design conditions in terms of linear matrix inequalities, which can be solved easily by using scientific software. To derive the design conditions, we use the generalized relaxation lemma (C. Briat, 2014). Moreover, we deal with conservatism which we often encounter in LMIs approaches. To this end, we derive the design conditions based on a descriptor form of control systems, which often provides some redundancies in derived conditions.

Applied System Innovation – Meen, Prior & Lam (Eds)
© 2016 Taylor & Francis Group, London, *ISBN 978-1-138-02893-7*

A framework for composing heterogeneous service tools involved in load testing lifecycle

Shin-Jie Lee
Computer and Network Center, National Cheng Kung University, Tainan City, Taiwan, R.O.C.
Department of Computer Science and Information Engineering, National Cheng Kung University, Tainan City, Taiwan, R.O.C.

You-Chen Lin & Kun-Hui Lin
Institute of Manufacturing Information and Systems, National Cheng Kung University, Tainan City, Taiwan, R.O.C.

Jie-Lin You
Department of Computer Science and Information Engineering, National Cheng Kung University, Tainan City, Taiwan, R.O.C.

ABSTRACT: Load testing is the process of applying ordinary stress to a software system to determine the system performance under normal conditions. In a typical load testing lifecycle, three kinds of service tools are involved: test case recording service tools that make testers easier to generate test cases through a web browsing-like behavior; test case execution service tools that exercise test cases with simulations of a large number of concurrent users; system resource monitoring service tools that provide information of system footprints during the test case executions. However, using these three kinds of service tools one by one to complete a load testing may require extra effort on operating and configuring each service. In this paper, we proposed a framework for composing the three types of service tools as an integrated service for load testing. A raw test case recorded by Badboy tool is automatically converted into an expanded test case that can be executed by JMeter. JMeter and Cacti are then automatically invoked by the framework. The execution time period of JMeter is automatically identified as the input to Cacti for resource monitoring of the system under test. The test report together with system footprints is also automatically generated. In the experimental evaluation, the result shows that the framework significantly save time on operating and configuring the load testing service tools than the traditional approach under a t-test.

Applied System Innovation – Meen, Prior & Lam (Eds)
© 2016 Taylor & Francis Group, London, ISBN 978-1-138-02893-7

A new particle swarm optimization technique with iterative learning control for high precision motion

Yi-Cheng Huang & Ming-Chi Hsu
Department of Mechatronics Engineering, National Changhua University of Education, Chaghua, Taiwan

ABSTRACT: This paper develops a new Improved Particle Swarm Optimization (IPSO) technique for adjusting the gains of PID controller, Iterative Learning Control (ILC) and the bandwidth of zero-phase Butterworth filter of the ILC. The conventional ILC learning process has the potential to excite rich frequency contents and to learn the error signals. However the learnable and unlearnable error signals should be separated for bettering control process along with the repetitions. Since producing a high frequency error condition should be avoided before the phase margin cause any trouble. Learnable error signals through a bandwidth tuning mechanism should be adaptively injected into learning control laws and thus reduce the tracking error effectively at every repetition. The filter bandwidth should be changed at every repetition for the shape of errors at frequency response thinking. Thus adaptive bandwidth in the ILC controller with the aid of IPSO tuning is proposed here. Simulation results show the new controller can cancel the errors efficiently as the process is repeated. Simulation results validate the effectiveness of the new IPSO-ILC for precision motion control.

Keywords: particle swarm optimization, iterative learning control, zero phase butterworth filter, empirical mode decomposition

Applied System Innovation – Meen, Prior & Lam (Eds)
© 2016 Taylor & Francis Group, London, ISBN 978-1-138-02893-7

Intelligent synchronization of uncertain chaotic systems using adaptive dynamic TSKCMAC

Ya-Fu Peng, Wen-Fong Hu & Yao-Jen Tsai
Department of Electrical Engineering, Chien Hsin University of Science and Technology, Taoyuan, Taiwan

ABSTRACT: In this paper, an adaptive control method based on dynamic Takagi-Sugeno-Kang type Cerebellar Model Articulation Controller (TSKCMAC) is developed for synchronizing uncertain chaotic system. This adaptive dynamic TSKCMAC system is composed of two parts. One is a dynamic TSKCMAC that is used to approximate an Ideal Backstepping Synchronization Control (IBSC) law, and the other is a compensated controller that is designed to compensate for the difference between the ideal controller and the dynamic TSKCMAC. The weights of the adaptive dynamic TSKCMAC system are on-line tuned by the derived adaptive law based on the Lyapunov synthesis and backstepping control technique, so that the stability of the system can be guaranteed. To check the theoretical results, the proposed controller is applied to control the Genesio chaotic system. Simulation results demonstrate the effectiveness of the proposed synchronization control scheme for chaotic dynamical systems with unknown dynamic functions.

Communication Network & Information Technology

Applied System Innovation – Meen, Prior & Lam (Eds)
© 2016 Taylor & Francis Group, London, ISBN 978-1-138-02893-7

The use of ZigBee wireless mesh networking for implementing smart homes

Chia-Hsin Cheng & Yeh-Wei Lin
Department of Electrical Engineering, National Formosa University, Huwei, Yunlin, Taiwan

Yung-Fa Huang
Department of Information and Communication Engineering, Chaoyang University of Technology, Wufeng, Taichung, Taiwan

ABSTRACT: This paper implements a Smart-Home system based on ZigBee Wireless Sensor Networks (WSNs) technology. The proposed system provides a monitoring system in the house. The monitor system uses an android Native Development Kit (NDK) and a Software Development Kit (SDK) to write an application interface for android mobile phones. In this system, we provide an interference to obtain sensor data and inquire into the node status. The use of mesh topology architecture is used to achieve a robust data transmission and upgrade the packet delivery ratio. Finally, the proposed system archives a reliable and efficient ZigBee mesh network.

Applied System Innovation – Meen, Prior & Lam (Eds)
© 2016 Taylor & Francis Group, London, ISBN 978-1-138-02893-7

The efficient one-time password authentication scheme for resisting off-line guessing attacks

Zhen-Yu Wu
Department of Information Management, National Penghu University of Science and Technology, Penghu, Taiwan

Chia-Hui Liu
Department of Digital Literature and Arts, St John's University, Taipei, Taiwan

Yu-Fang Chung
Department of Electrical Engineering, Tunghai University, Taichung, Taiwan

Tsung-Chih Hsiao
College of Computer Science and Technology, Huaqiao University, Fujian, China

ABSTRACT: User authentication protocols ensure the security of user communication and data transmission over insecure networks. Among the various current authenticated mechanisms, the password-based user authentication, because of its convenience and efficiency, is the most widely employed mechanism in electronic applications. Even though the password is endowed with the advantageous properties of simplicity and human memory, it can easily succumb to attacks that employ brute force, such as offline guessing attacks that various existing schemes suffer from, or to spoofing and impersonation problems that occur once the password is hacked. Therefore, this paper intends to investigate the dynamic password-based user authentication scheme, where the characteristic of a dynamic password disables attackers from catching the correct password each time and prevent the attackers from guessing the users' passwords. Besides, this scheme can also resist common attacks, such as replay attacks, stolen-verifier attacks, server spoofing attacks, and impersonation attacks, among others.

Keywords: Data transmission, Password-based user authentication, Off-line guessing attacks, Dynamic password

Applied System Innovation – Meen, Prior & Lam (Eds)
© 2016 Taylor & Francis Group, London, ISBN 978-1-138-02893-7

An adaptation scheme to ambient light interference for Visible Light Communication link

Cheng-Han Li & Jenn-Kaie Lain
Department of Electronic Engineering, YunTech, Yunlin, Taiwan

ABSTRACT: A joint Pulse-Width Modulation and Non-Return-to-Zero On-Off Keying (PWM/NRZ-OOK) is investigated as a physical-layer signal format for the Visible Light Communication (VLC) systems. Being the underlying PWM architecture of the NRZ-OOK transmissions, the joint PWM/NRZ-OOK increases bandwidth utilization while maintaining dimming control of light emitting diodes in VLC. In order to reduce the effect of ambient light interference, a decision threshold adaptation scheme is proposed and evaluated. Implementation results demonstrated that the proposed decision threshold adaptation scheme is simple and effective when VLC is operating in an environment with ambient light interference.

Applied System Innovation – Meen, Prior & Lam (Eds)
© 2016 Taylor & Francis Group, London, ISBN 978-1-138-02893-7

User acceptance of knowledge management on-line interactive system used in architectural design learning design process

Yun-Wu Wu, Chin-Min Chen & Kuo-Hua Weng
Department of Architecture, China University of Technology, Taiwan

ABSTRACT: The purpose of this study is to develop a knowledge management on-line interactive learning system for architectural design learning, helping students to share, extract, use and create their design knowledge through web-based interactive learning activities based on the process of socialization, combination, externalization and internalization other than learning in the "design studio" at school. In addition, the technology acceptance model is used in this study to measure how students accept the system developed in this study by analyzing the user behavioral and environmental factors that can affect their use of the system. Then the TAM questionnaire survey results indicate that the digital on-line interactive learning environment based on knowledge management developed in this study can effectively reduce the challenge of cognitive overload problems for students and help to promote better learning results. Therefore, it can be concluded that the knowledge management on-line interactive learning system developed in this study is very helpful for learning of architectural design.

Keywords: Internet of Things; ZigBee technology; intelligent living system; wireless.

Green and High Performance System Technology

Applied System Innovation – Meen, Prior & Lam (Eds)
© 2016 Taylor & Francis Group, London, ISBN 978-1-138-02893-7

Application of 3D printing to the cold forging tool system of long hex flange nuts

Shao-Yi Hsia
Department of Mechanical and Automation Engineering, Kao Yuan University, Kaohsiung City, Taiwan, R.O.C.

Yu-Tuan Chou
Department of Applied Geoinformatics, Chia Nan University of Pharmacy and Science, Tainan City, Taiwan, R.O.C.

ABSTRACT: Cold forging has played a critical role in fasteners and has been applied to the automobile industry, construction industry, aerospace industry, and for living products, cold forging presents opportunities for manufacturing more products. By using computer simulation, this study attempts to analyze the process of creating machine parts, such as special nuts. The DEFORM-3D forming software was applied to analyze the process at various stages in the computer simulation, and the compression test was used for the flow stress equation, to reach the real plastic forming behavior. At the same time, the 3D printing technology was also utilized to create the cold forging dies and the deformed workpiece. It allows engineers to gain a better understanding of the tooling design and development phase, and touches a part that was previously just a simulation result from the DEFORM-3D forming software. The research results will benefit the machinery businesses in realizing the forging load and forming conditions at various stages before the fastener formation. In addition, to planning a proper die design and production; the quality of the produced long hex flange nuts would be more stable to promote industrial competitiveness.

Applied System Innovation – Meen, Prior & Lam (Eds)
© 2016 Taylor & Francis Group, London, *ISBN 978-1-138-02893-7*

Assessment of visual comfort at leisure space using the simulation process

Bi-Wen Lee
Department of Tourism Management, Shih Chien University, Kaohsiung City, Taiwan

Yu-Tuan Chou & Hsin-Yi Shih
Department of Applied Geoinformatics, Chia Nan University of Pharmacy and Science, Tainan City, Taiwan

ABSTRACT: There is a high relevance between visual comfort and lighting design in indoor spaces. Lighting layout is crucial to workplaces and living environment, especially at leisure spaces. Heretofore, indoor visual design has been processed by experience or by using a simple computing formula, and just obtained one or two indices for evaluation. The method and result is not sufficient for high-quality building design nowadays. This article is aimed to evaluate the visual comfort at leisure spaces using the raytrace and radiosity simulation technique to analyze the direct and indirect illuminations of different lighting layouts at a leisure space, based on a digital three-dimensional indoor model. The commercial software of Autodesk 3ds Max Design is applied to conduct a three-dimensional leisure space environment model with materials. The illuminance simulation of single artificial lighting is conducted first, and the experimental verification of simulation error is under 10%. Then, luminance distribution, uniformity and glare index are calculated through the illuminance of various areas gained from the multi-artificial lighting. The results also indicate that the illuminance and visual comfort are not enough in some leisure spaces. Therefore, designers can acquire more adequate information for evaluating and designing light layouts in the preliminary design stage via the efficient simulation process.

Applied System Innovation – Meen, Prior & Lam (Eds)
© 2016 Taylor & Francis Group, London, ISBN 978-1-138-02893-7

Design improvement for acoustic comfort of a small indoor space

H.Y. Shih, Y.T. Chou & P.Y. Lai
Chia Nan University of Pharmacy and Science, Tainan City, Taiwan

S.Y. Hsia
Kao Yuan University, Kaohsiung City, Taiwan

ABSTRACT: Recently, acoustic characteristics of the indoor space have become more important due to society's economic development needs. For sustainable and easy use features, container houses are gradually widely used in quite a lot of space applications, such as temporary premises, office and exhibition space. They still need to remain as a pleasant and quiet sound environment. In this paper, green building analysis software Ecotect Analysis is used to calculate the sound characteristics of the container house. And the decorated material of this small indoor space is redesigned for obtaining the acoustic comfort at the optimal reverberation time. First, a three-dimensional model of the container house is constructed by default tools of software. The indoor acoustic characteristics for various design conditions can be obtained by the simulation process. The results indicated that the reverberation time distribution of the original container houses is 140–315 ms. After changing the interior material, its reverberation time distribution is 160–680 ms. For less than 1 kHz frequency, the reverberation time is located within the optimal range of acoustic comfort. Following the design process, the spatial designers can assess the indoor acoustic characteristics in the concept design stage. By changing the interior materials, the acoustic comfort environment of the building is derived. Meanwhile, the quality of living can be improved and reduction in construction costs can be implemented.

Applied System Innovation – Meen, Prior & Lam (Eds)
© 2016 Taylor & Francis Group, London, ISBN 978-1-138-02893-7

Enhancement on 3D model construction of corroded pipelines using the parametric design approach

Yu-Tuan Chou
Department of Applied Geoinformatics, Chia Nan University of Pharmacy and Science, Tainan City, Taiwan

Shao-Yi Hsia
Department of Mechanical and Automation Engineering, Kao Yuan University, Kaohsiung City, Taiwan

Ting-Wei Chang
Department of Applied Geoinformatics, Chia Nan University of Pharmacy and Science, Tainan City, Taiwan

ABSTRACT: For the petrochemical industry, the serious accidents are always induced from the leakage of corroded pipelines. The component detection, service assessment and engineering management of transportation elements are most important for assuring the plant operation. Recently, API-579, simultaneously containing the conservative and predictable assessment process, is gradually replaced by the traditional check process. However, the accurate stress distribution of corroded components should be analyzed before introducing the assessment criteria. In this paper, the enhanced approach for the 3D parametric model construction of corroded pipelines is developed by integrating the SolidWorks Application Program Interface (API) and database system. First, the attribute of the component, such as size and material properties, can be obtained from the component database. And the 3D model of the pipe is built using SolidWorks API language. Next, the detection data for thinning of corrosion are obtained from the ultrasonic nondestructive method. The corroded 3D model can be constructed from the obtained data. Combining the two models by using the Boolean operation, the 3D model of corroded pipeline can be derived and imported into the process of finite element analysis. Two case studies, surface model and solid model for the local metal loss of pipelines, are introduced to verify the feasibility of the approach. It is shown that the operation time of the new approach is only one-tenth of the former. The research results would benefit the assessment process of petrochemical plant maintenance and increase the operation efficiency.

Applied System Innovation – Meen, Prior & Lam (Eds)
© 2016 Taylor & Francis Group, London, ISBN 978-1-138-02893-7

Resource distribution and utilization technologies of renewable energy in Hubei province, China

Y.P. Xiao, Q.K. Wang, T.T. Mei & J.L. Zhou
Wuhan University of Technology, Wuhan, Hubei, China

ABSTRACT: Lacking coal, oil, and natural gas, Hubei province is eager to vigorously promote the development and application of renewable energy. Thus, on the background of the renewable energy use policy, combined with the "twelfth five-year" plan of energy development in Hubei province, the actively utilized technologies of renewable energy, including hydroelectric technology, wind power technology, utilization technology of solar energy, utilization of biomass energy, and ground source heat pump(GSHP) technology are discussed in this paper on the basis of a large number of original statistic data of the current situation of renewable energy resource distribution. Besides, according to large amounts of original stats as well, the economic potential of renewable energy is also confirmed by analyzing the energy utilization status. And eventually, the utilization situation of renewable energy is revealed, which aims at providing theoretical references for promoting the development of renewable energy technology in Hubei province of China.

Keywords: renewable energy; resource distribution; hubei province

Advanced Dynamics and Vibration Technology for Engineering Applications

Applied System Innovation – Meen, Prior & Lam (Eds)
© 2016 Taylor & Francis Group, London, ISBN 978-1-138-02893-7

A highly dynamic current loop design for Permanent Magnet Synchronous Motor drive system

Chiu-Keng Lai & Yaw-Ting Tsao
National Chin-Yi University of Technology, Taichung, Taiwan, R.O.C.

ABSTRACT: To improve the dynamic responses of motor drive system, developing an inner current loop with fast response is necessary. In this paper, a wide bandwidth current loop system for Permanent Magnet Synchronous Motor (PMSM) drive using FPGA-based controller is present. Most of the controllers for vector control are realized by Digital Signal Processors (DSPs) due to the complete and compact hardware functions of DSP, such as SVPWM, A/D converter, Quadrature Encoder Counter (QEP), and other functions for motor control. However, the drive systems realized by DSP are limited by the hardware structure, and hardly reach an enough fast response. In this paper, we propose an FPGA realized current loop system which considers the time delay of phase current sensor and the processing time of digital controller. Simulation and practical experimental results are used to verify the performance.

Applied System Innovation – Meen, Prior & Lam (Eds)
© 2016 Taylor & Francis Group, London, ISBN 978-1-138-02893-7

The effect of strain rate on the response of nonlinear material models of concrete

P. Kral, P. Hradil & J. Kala
Faculty of Civil Engineering, Brno University of Technology, Czech Republic

ABSTRACT: The subject of this contribution is the testing of the effect of strain rate on the response of nonlinear material models during compression and tensile loading at higher speeds than that of quasi-static loading. The results gained from numerical simulations during loading at higher speeds are compared with results gained from numerical simulations of quasi-static loading.

Applied System Innovation – Meen, Prior & Lam (Eds)
© 2016 Taylor & Francis Group, London, ISBN 978-1-138-02893-7

CFD analysis of membrane structure

J. Kala
Department of Structural Mechanics, Faculty of Civil Engineering,
Brno University of Technology, Brno, Czech Republic

ABSTRACT: Textile membrane structures are one of the most popular elements of modern architecture. These structures consist of textile membranes, which are tensioned by ropes or are attached to a solid frame. Due to this, it is possible to assemble structures of various shapes, de facto without limits. Nevertheless, these structures are more vulnerable to the effects of wind flow. This paper deals with the effects of wind on these structures. Because the shape of these structures is mostly variable, in such cases it is not possible to determine the wind load by the standards. Therefore, it is of suitable use to determine the effects of wind CFD (Computational Fluid Dynamics). The influence of the shape and structure is observed in this paper in terms of statics and dynamics. Further the character of the dynamic component is observed in terms of wind flow. The software system ANSYS/CFX was used for this analysis.

Applied System Innovation – Meen, Prior & Lam (Eds)
© 2016 Taylor & Francis Group, London, ISBN 978-1-138-02893-7

Numerical analysis of dynamic response in railway switches and crossings

V. Salajka
Faculty of Civil Engineering, Brno University of Technology, Brno, Czech Republic

M. Smolka
DT - Vyhybkarna a strojirna, a.s., Prostejov, Czech Republic

O. Plasek & J. Kala
Faculty of Civil Engineering, Brno University of Technology, Brno, Czech Republic

ABSTRACT: Extreme stress in all components of a permanent railway track occurs in switches and crossings during passage of railway vehicles. Additional dynamic loading originates from changes in track stiffness along the railway turnout or alternatively impacts the forces caused at points of geometric imperfections. The stress can be reduced by controlling track stiffness via rail fastening with a different elasticity of rail pads or special elastic pads in slide plates in switches. A procedure for numerical analysis of the dynamic response during the passage of railway vehicles is described. The solution is based on Finite Element Method (FEM), which is used for calculation of track stresses. The FEM model with a fine structure comprises all the components of switches and crossing including movable parts. The excitation forces are defined on the basis of the supposed interaction between the track and vehicle. The track stiffness defined by FEM analyses is used for the calculation of dynamic vertical and lateral wheel load. The special model of the railway vehicle was built to calculate forces at points of abrupt stiffness changes and geometric imperfections at the frog structure. The excitation forces obtained are applied backwards in the dynamic responses calculation by FEM. The analyses described above were used as a tool for designing track stiffness, controlled by rail pad elasticity. The final arrangement of rail fastening in switch and crossing of a typical turnout structure was found by the optimization of dynamic response within the interaction between the railway track and vehicle.

Intelligent Algorithms, Systems and Applications

Applied System Innovation – Meen, Prior & Lam (Eds)
© 2016 Taylor & Francis Group, London, *ISBN 978-1-138-02893-7*

Liveness face detection based on single image

Chin Lun Lai & Chiu Yuan Tai
Department of Communication Engineering, Oriental Institute of Technology, Taipei, Taiwan

ABSTRACT: An effective and efficient algorithm is proposed in this paper to distinguish the live face from fake face images in the bioinformatics authentication systems to improve the system reliability. By analyzing the spectrum distribution of the interested face region, fake faces which shown in LCD displays can be easily picked out from captured live faces. The good performance of the simulation results show that the proposed method is practical and is suitable to be implemented in live face recognition systems.

Applied System Innovation – Meen, Prior & Lam (Eds)
© 2016 Taylor & Francis Group, London, *ISBN 978-1-138-02893-7*

A wavelength-switchable laser source for dense wavelength-division multiplexing optical communication system

Yan Ju Chiang
Department of Electronics Engineering, Oriental Institute of Technology, New Taipei City, Taiwan, R.O.C.

ABSTRACT: A multi wavelength-switchable Erbium Doped Fiber Laser (EDFL), employing fiber Bragg gratings, is proposed in this paper. The experimental results demonstrate that the proposed EDFL has advantages such as moderate power with high stability, wide switchable range, high side-mode suppression ratio, and a narrow separation between two switchable wavelengths *(0.15 nm)*.

Keywords: erbium-doped fiber laser; fiber bragg grating; tunable laser

Applied System Innovation – Meen, Prior & Lam (Eds)
© 2016 Taylor & Francis Group, London, ISBN 978-1-138-02893-7

Active shaping of deformable object using robotic manipulator

Kang Hyun Nam
Department of Mechanical Engineering, Yeoungnam University, Gyeongsan, Korea

Sang-Ryong Lee, CheolWoo Park & Choon-Young Lee
Department of Mechanical Engineering, Kyungpook National University, Gyeongbuk, Korea

ABSTRACT: Physical modeling of deformable object is used in the simulation of virtual world with soft materials. The properties of dough in food industry determine the quality of major products like breads, noodles, and pasta for human consumption. The lattice structure of rheological elements is chosen to form wheat dough model in the simulation of active shaping of the deformation. The force-dependent damper plays a critical role in the plastic deformation, which shows nonlinear transfer function. In this paper, we modeled the viscoelastic objects with monotonically decreasing nonlinear damper as piece-wise linear model. The manipulation of robotic hand to form deformation is conducted through mechanical simulation to show dexterous dynamic shaping of flour dough.

Applied System Innovation – Meen, Prior & Lam (Eds)
© 2016 Taylor & Francis Group, London, *ISBN 978-1-138-02893-7*

Competition-based Differential Evolution for global numerical optimization

Chun-Ling Lin
Department of Electrical Engineering, Ming Chi University of Technology, New Taipei City, Taiwan, R.O.C.

Huang-Lyu Wu & Sheng-Ta Hsieh
Department of Communication Engineering, Oriental Institute of Technology, New Taipei City, Taiwan, R.O.C.

ABSTRACT: Differential Evolution (DE) is arguably one of the most powerful stochastic real-parameter optimization algorithms in the mainstream. In this paper, a competition-based mutation strategy is proposed for enhancing DE's solution searching ability. It can efficiently produce useful vectors to guide vectors to perform a wide search. A similar concept is also applied on crossover. It can speed up convergence and perform a deep search. In the experiments, 15 CEC 2005 test functions, which include five uni-modal and ten multi-modal functions, are selected to verify the performance of proposed method and compare it with five recent DE variants. From the results, it can be observed that the proposed method exhibits better results than other related works.

Applied System Innovation – Meen, Prior & Lam (Eds)
© 2016 Taylor & Francis Group, London, *ISBN 978-1-138-02893-7*

Artificial Bee Colony with enhanced search strategies for food sources

Sheng-Ta Hsieh
Department of Communication Engineering, Oriental Institute of Technology, New Taipei City, Taiwan, R.O.C.

Chun-Ling Lin
Department of Electrical Engineering, Ming Chi University of Technology, New Taipei City, Taiwan, R.O.C.

Shih-Yuan Chiu
Systems Development Center, Chung-Shan Institute of Science and Technology, Taoyuan City, Taiwan, R.O.C.

ABSTRACT: Artificial Bee Colony (ABC) is a population-based optimizer. It simulates bees' foraging behavior for finding better food sources (solutions). In order to improve bees' searching ability, in this paper, the moving behavior of onlooker bees is modified. Also, the population manager is involved to produce potential bees and eliminate redundant bees according to the solution searching status. The bees which are generated by the proposed method can guide other bees toward the potential solution space and easier to find better solutions. In order to test the efficiency of the proposed method and compare it with other ABC variants, eleven test functions of CEC 2005 are adopted. It can be observed from the results, that the proposed method performs better than other works on most test functions.

Applied System Innovation – Meen, Prior & Lam (Eds)
© 2016 Taylor & Francis Group, London, *ISBN 978-1-138-02893-7*

Mean-shift clustering of SIFT keypoints extracted from repetitive structures by considering both gradient and location

Sunmin Lee & Yong C. Kim
Department of Electrical and Computer Engineering, University of Seoul, South Korea

ABSTRACT: We propose a two-step clustering of SIFT keypoints extracted from urban buildings. Windows and walls of modern buildings are so repetitively patterned that the descriptor vectors for such repetitive keypoints are very similar to each other. Hence, similarity based matching between keypoints is highly prone to error. This problem can be alleviated by cluster-to-cluster matching. Mean-shift clustering groups similar keypoints into a single cluster and so the ambiguity in matching between keypoints significantly decreases. Further improvement can be obtained by relaxation of matching such that the structural consistency among matched clusters is preserved between two images. We tested the proposed scheme on several images of buildings. Significant improvement in terms of precision rate and recall rate are achieved.

Service Design Essentials and Practices

Applied System Innovation – Meen, Prior & Lam (Eds)
© 2016 Taylor & Francis Group, London, ISBN 978-1-138-02893-7

Analysis of the service design of aboriginal tribes' experience activities—taking an example of the Laiji Tribe in Alishan township

Shyh-Huei Hwang & Hsiu-Ting Su
Graduate School of Design, National Yunlin University of Science and Technology, Yunlin, Taiwan

ABSTRACT: Following the enhancement of aborigines' self-awareness over the last few years, tribal experience activities featured with local historical, cultural and natural environmental elements have not only become an important direction for the tribal experience economy, but also made relevant cases and researches hot topics. Nevertheless, there are still some limitations for the analysis of tribal experience activities from the perspective of service design. Laiji Tribe in Alishan Township is a typical case as its tribal experience is integrated with features of local traditional culture. Having Laiji Tribe as the research target, this research has adopted in-depth interview and observation methods to acquire relevant research data in order to discuss Laiji Tribe's customer journey maps and to analyze the service design of its craft experience. The results indicate that: 1) the service design of tribal experience activities mainly connects with the tribe's original primary agriculture and forestry industries to develop various abundant tribal experiences such as farming experience, eco-tourism, creative crafts, tribal tour, guesthouse experience etc.; 2) tribal craft experiences enable customers to get involved with the core connotation of aboriginal people's knowledge and techniques such as their history, culture and traditional crafts. In enables customers to extend their experiences from the service of substantial products to emotions; 3) the design of tribal craft experiences is to have co-creations and exchanges between the customers and people from the tribe. Apart from providing customers abundant cultural stories, tribal people also reflect the sustainability issue of the operation of its traditional culture; 4) intangible cultural connotations plus crafts that can be made with local materials may attract customers to tribes lacked of exhibition and sales hardware facilities for having some craft experiences. The research results may be used as the foundation of the following research on the service design of tribal craft experience.

Applied System Innovation – Meen, Prior & Lam (Eds)
© 2016 Taylor & Francis Group, London, ISBN 978-1-138-02893-7

Co-creation design practice in intergenerational gaming product

Chia-Ling Chang
Department of Creative Product Design and Management, Far East University, Tainan, Taiwan

Wang-Chin Tsai
Department of Product and Media Design, Fo-Guang University, Yilan, Taiwan

Ding-Bang Luh
Department of Industrial Design, National Cheng Kung University, Tainan, Taiwan

ABSTRACT: In response to the aging society and the change in family's economic structure, the phenomenon of cross-generational parenting is growing gradually, leading to grandparents and grandchildren (the elderly and children) spending time and playing together more often. A product suitable for both becomes an important medium for physical and mental interaction between them, but little research has taken into account the elderly and children's physical and mental condition, let alone products made for both groups and that can facilitate creative interaction in between. Based on the idea of "incorporating user creativity", the study targets the elderly and children as users to develop cross-age products, through which the grandparents and grandchildren can strengthen their physical and creative development and optimize their sensory coordination when playing games. This outcome won the Red Dot Award: Design Concept 2014 and the silver medal in Seoul International Invention Fair in 2014, showing that its design accommodates both professional expectation and social trend.

Applied System Innovation – Meen, Prior & Lam (Eds)
© 2016 Taylor & Francis Group, London, ISBN 978-1-138-02893-7

Positive creativity as a mediator for facilitating goal congruence among team members in a design project

Wu Chi-Hua & Luh Ding-Bang
National Cheng Kung University, Tainan, Taiwan

ABSTRACT: In this paper, a goal represents an imagination toward the future. It provides directions and motivation to enable communication and negotiation among teams and team members to transform efficiently, to adapt to the outer environment, and to reflect on inner demands. Goals are important for both team and team members. However, since every team is composed of more than one individual, the conflicts between an individual's purpose and team goals are inevitable, especially for design teams. Design, by essence is to 'create something out of nothing', requiring the designers' dedication to produce innovative solutions. Under this circumstance, incongruence between the individual and team goals may increase the communication cost and decreases the individual's willingness to innovate, causing loss of innovation efficacy.

In the last few decades, the importance of 'goal congruence' has been continuously emphasized and goal conflicts are well-documented and studied in organizational and psychological researches. Nonetheless, feasible ways of cooperation are yet to be provided to effectively integrate personal dedication and team goal achievement. 'Positive Creativity' provides a mechanism to foresee all ideal visions from the current standpoint and enables team members to select designs that satisfy both the group target and personal purposes, and bares the potential of enhancing goal congruence.

Applied System Innovation – Meen, Prior & Lam (Eds)
© 2016 Taylor & Francis Group, London, ISBN 978-1-138-02893-7

Analysis of smartphone OS interface typology with Mobile Information Architecture

Ti-Wei Chang & Ding-Bang Luh
Department of Industrial Design, National Cheng Kung University, Tainan, Taiwan

ABSTRACT: The purpose of this study was to investigate the user preference of interface typology in the smartphone operating system, based on the aspect of the Information Architecture (IA). Relevant studies focus on the World Wide Web and applications. The concept that introduced into mobile information devices enhances diversity, intuition, and flexibility in interface typology. Thus, it is necessary to re-examine interface typology with Mobile Information Architecture (MIA), which consists of four aspects, namely content, structure, navigation, and representation. In some cases, users interact with the OS interface in two ways: either they adapt MIA and maintain the OS interface, or they try to amend the interface to suit their habits. Based on MIA with a semi-structured interview, this study investigated 11 expert users for their experiences in interface design, operating inclination, and adaptability. It is expected to suggest design improvement strategies in accordance with adaptation analysis based on four aspects of the MIA. The result of this study is to provide a more comprehensive interface typology with MIA, and to propose guidelines for adaptation design for higher system usability and efficiency.

Applied System Innovation – Meen, Prior & Lam (Eds)
© 2016 Taylor & Francis Group, London, ISBN 978-1-138-02893-7

Aspects and factors analyses of public service design for pedestrian environment

Hsing-Yu Hsu & Ding-Bang Luh
Department of Industrial Design, National Cheng Kung University, Tainan, Taiwan

ABSTRACT: The essence of the public service design is to allow most users with uncertainty to obtain fine experiences and benefits when encountering public resources and services. However, most service design practices are currently applied to commercial businesses, giving little attention to public benefit. This is evident in pedestrian environments nowadays in many cities. The pedestrian environment is not only a place for walking, but also a space to set up street public facilities. However, the planning and evaluation of the pedestrian environment is far less considered from a pedestrian's perspective. An example is the conflicts of the space between the pedestrian environment against the transformer box and the drop between the arcades, where many bicycles and pedestrians are on the same path.

In order to conduct a public service design for the pedestrian environment, it is fundamental to get insight into the needs from at least two perspectives, i.e., that of the pedestrian and that of the city government. This study aims to analyze the major aspects and factors that should be considered in designing a pedestrian environment from the minority's point of view. A minority-centered design is used to reveal problems and suggestions for improvements in the current design of pedestrian environment.

Keywords: public service design; pedestrian environment; minority-centered design

Applied System Innovation – Meen, Prior & Lam (Eds)
© 2016 Taylor & Francis Group, London, ISBN 978-1-138-02893-7

Author index